BRITISH GEOLOGICAL SURVEY

J PATTISON
N G BERRIDGE
J M ALLSOP
I P WILKINSON

Geology of the country around Sudbury (Suffolk)

CONTRIBUTORS

Stratigraphy
R A Ellison
D Millward
A Smith

Economic geology
P M Hopson
N S Robins

Biostratigraphy
A A Morter
C J Wood

Petrology
R J Merriman

Memoir for 1:50 000 geological sheet 206
(England and Wales)

LONDON: HMSO   1993

ISBN 0 11 884500 4

Bibliographical reference
PATTISON, J, BERRIDGE, N G, ALLSOP, J M, AND WILKINSON, I P. 1993. Geology of the country around Sudbury (Suffolk). *Memoir of the British Geological Survey*, sheet 206 (England and Wales).

Authors
J Pattison, MSc
N G Berridge, BSc, PhD
J M Allsop, BSc
I P Wilkinson, MSc, PhD
British Geological Survey, Keyworth

Contributors
R A Ellison, BSc
P M Hopson, BSc
R J Merriman, BSc
A Smith, BSc
British Geological Survey, Keyworth

D Millward, BSc, PhD
British Geological Survey, Newcastle upon Tyne

N S Robins, MSc
British Geological Survey, Wallingford

A A Morter, BSc
C J Wood, BSc
formerly British Geological Survey

Other publications of the Survey dealing with this and adjoining districts (see also pp. x and 55).

BOOKS
Memoirs
Cambridge (188), 1969
Bury St Edmunds (189), 1990
Great Dunmow (222), 1990
Braintree (223), 1986

British Regional Geology
London and the Thames Valley (4th edition), in preparation
East Anglia (4th edition), 1961

Well catalogue
Records of wells in the area of New Series One-Inch Sudbury (206) sheet, 1965

MAPS
1:625 000
Solid geology (south sheet)
Quaternary geology (south sheet)
Aeromagnetic (south sheet)

1:250 000
East Anglia (Sheet 52N 00)
  Solid geology, 1985
  Quaternary and Sea bed sediments, 1989
  Aeromagnetic anomaly, 1982
  Bouguer gravity anomaly, 1981

1:125 000
Hydrogeological map of Southern East Anglia (two sheets), 1981
(in collaboration with the Anglian Water Authority)

1:50 000 and 1:63 350 (Solid and Drift)
Cambridge (188), 1981
Bury St Edmunds (189), 1982
Saffron Walden (205), 1952
Ipswich (207), 1990
Great Dunmow (222), 1990
Braintree (223), 1982

Printed in the UK for HMSO
Dd 0292043 12/93 C8 531/3 12521

Geology of the country around Sudbury (Suffolk)

The Sudbury district is quintessential East Anglia, with rolling farmland, half-timbered or pink-washed cottages and ornate flint-built churches. It includes the highest parts of the East Anglian plateau and much of the beautiful Stour valley, which dissects that plateau along the Suffolk/Essex border. The district is blanketed by a thick cover of glacial clay till, which is significantly breached only by the deeper valleys, in which older drift deposits and the underlying solid rocks are exposed. The older drift includes sands and gravels of both glacial and fluviatile origin; the latter, which are considered to be products of a periglacial proto-Thames river, constitute a major bulk-aggregate resource.

The youngest solid rocks are spreads of marine sands and gravels laid down in the late Pliocene to early Pleistocene 'Crag' sea. In the south-eastern areas, the drift and Crag partly overlie Tertiary rocks, which dip gently south-south-eastwards. Underlying the whole district is a great thickness of Chalk, an important source of groundwater, but previously little studied. A broad picture of the local Chalk stratigraphy and structure has now been deduced using palaeontology and borehole geophysical logs.

A deep borehole at Clare has proved, beneath the Chalk, thin Gault clay overlying Silurian rocks. It also confirmed the remarkable depth of the drift-filled buried channel which lies under the modern Stour valley, the deepest of many in East Anglia, and thought to have been gouged out by subglacial meltwater torrents.

Cover photograph
The Old School, Sudbury. (A 13258)

Chalk, Thanet Beds, Crag and Till in an old pit in Sudbury [8734 4213].

The Chalk is succeeded by thin basal Thanet Beds, consisting of rounded and nodular glauconite-coated flints (the Bullhead Bed) and blocky pinkish fine sand. The top few centimetres of the Thanet Beds are brecciated and overlain by bedded, rusty brown, medium-grained to coarse sand (Crag) with local calcrete. Chalky till with basal calcretes caps the section [A 13265].

CONTENTS

One   Introduction   1
History of research   3
Outline of geological history   4

Two   Pre-Cretaceous rocks   6
Pre-Silurian   6
Silurian   6
Devonian   9
Triassic   10

Three   Lower Cretaceous: Gault   12
Details   14

Four   Upper Cretaceous: Chalk (including Cambridge Greensand)   15
Cambridge Greensand   15
Cambridge Greensand to base of Middle Chalk   18
  Porcellanous Beds   18
  Chalk Marl above the Porcellanous Beds   18
  Lower Chalk above the Chalk Marl   20
Middle Chalk   20
Upper Chalk   21
  Turonian to Coniacian   21
  Early Santonian   22
  Late Santonian   22
  Campanian   23

Five   Palaeocene and Eocene: 'Lower London Tertiaries' and London Clay   24
Thanet Beds   25
  Details   27
Woolwich and Reading Beds   27
  Details   28
'Lower London Tertiaries', undifferentiated   28
  Details   28
London Clay   29
  Details   31

Six   Pliocene–Pleistocene Solid: Crag   33
  Details   33

Seven   Structure   35
Pre-Mesozoic structures   35
Mesozoic and Post-Mesozoic structures   36

Eight   Pre-Anglian drift deposits: Kesgrave Sands and Gravels   37
  Details   37

Nine   Anglian drift deposits   38
Drift-filled channels   39
  Details   40
Till   42
  Details   44
Glacial Sand and Gravel   45
  Details   47
Glacial Silt   48
  Details   48

Ten   Post-Anglian drift deposits   49
Lacustrine deposits   49
Head   49
  Details   50
River Terrace Deposits   51
  Details   52
Alluvium   52
  Details   52
Peat   53
Landslip   53

Eleven   Economic geology   54
Chalk   54
Sand and gravel   55
Brick clay   57
Hydrogeology and water supply   58
Ground instability   59

References   61

Appendices   65
1   Key boreholes and sections   65
2   BGS photographs   66

Fossil index   67

General index   69

FIGURES

1   Sketch map showing the physical features of the district   1
2   Simplified map of the solid geology   3
3   Bouguer gravity anomaly map of the Sudbury region   7
4   Aeromagnetic anomaly map of the Sudbury region   7
5   Gamma–sonic crossplot derived from downhole geophysical logs for the Clare Borehole   8
6   Histograms for the (a) sonic (velocity), (b) resistivity, (c) density and (d) gamma-ray logs for the Silurian rocks in the Clare Borehole   9
7   Sub-Mesozoic geology of the Sudbury region   10
8   Stratigraphy of the Gault in the Clare Borehole   12
9   Correlation of Gault sequences proved in East Anglia   13
10   Chalk localities and structure contours on the Chalk Rock   16
11   Inferred lithostratigraphy and tentative biostratigraphical and chronostratigraphical classification of the Chalk in the district   17
12   Biostratigraphical classification and distribution of selected macro- and microfossils in the Lower Chalk of the Clare Borehole   19

13 Correlation of 16-inch normal resistivity logs in the Middle and Upper Chalk for selected boreholes along the middle Stour valley and farther east 21
14 Contours on the base of the Palaeogene strata 24
15 Generalised isopachyte map of the Palaeogene strata 25
16 Diagrammatic section showing inferred stratigraphical relationships along an approximately north to south line in the east-central part of the district (TL 94 NW and SW) 34
17 Conjectural geological structures in the 'basement' rocks of the Sudbury region 35
18 Diagrammatic section showing spatial relationships between the different Anglian glacigenic deposits in the south-central parts of the district 38
19 Rockhead contours in the district 40
20 Borehole sections along the Stour tunnel valley 41
21 Generalised thickness of till 42
22 Generalised contours on the base of the till 43
23 Plan and simplified cross-section of Cornard brick pit 50
24 Plan showing geographical cover of Mineral Assessment Reports in the district 55
25 Summary assessment of sand- and gravel-bearing deposits in the district 56
26 Diagram to show the descriptive categories used in the classification of sand and gravel 57

PLATES

Frontispiece Chalk, Thanet Beds, Crag and till in an old pit in Sudbury
1 A valley cut into the west Suffolk till plateau at Hartest 2
2 Junction between Thanet Beds and Upper Chalk in an old pit at Ballingdon 26
3 London Clay–Woolwich and Reading Beds junction at Cornard brick pit 29
4 London Clay with ash bands in Bulmer brick pit 30
5 Fossiliferous cementstone bed in London Clay at Bulmer brick pit 31
6 Banded till at Edwardstone gravel pit 45
7 Banded till at Bear's Pit, Acton 46
8 Well-bedded Glacial Sand and Gravel in a pit north-north-east of Glemsford 47
9 The parish church at Long Melford with largely flint-built walls 54

TABLES

1 Physical properties of Mesozoic and basement rocks in southern East Anglia derived from analysis of the Clare Borehole geophysical logs and other sources 8
2 Chronostratigraphical and lithostratigraphical classification of Palaeogene strata 25
3 Major sand and gravel operations 57
4 Results of pumping tests on the Chalk aquifer 58
5 Typical chemical analyses of borehole waters from the Chalk 59

PREFACE

The district described in this memoir is typical of prosperous, rural East Anglia. The local economy and landscape are dominated by agriculture. The glacial 'chalky boulder clay', which covers at least two thirds of the district, weathers to form brown loam soils and good, mostly Grade Two, agricultural land from which wheat is the main crop. Geology, at first glance, appears to be of minor importance as an influence on the scenery and on industry other than agriculture. However, it is the primary control on the nature and distribution of two major resources, aggregate and groundwater. The boulder clay both partly incorporates and largely overlies several extensive sand and gravel formations which constitute a major aggregate resource for south-east England. An assessment of these sand and gravel deposits by BGS provided the initial impetus for the geological resurvey described herein. The second geology-related local resource, groundwater in the Chalk aquifer, is used to augment river abstraction. The structure of the Chalk, which has a major influence on groundwater movement, has been determined using borehole geophysical logs. Other resources in the area include brick clay in the London Clay.

The gently undulating land surface gives no hint of the considerably greater relief on the concealed surface underlying the glacial deposits. Several deep 'tunnel-valleys', are cut into the top of the Chalk, one extending down to at least 112 m below sea level, below the line of the Stour valley between Wixoe and Long Melford. The joint Anglian Water Authority/BGS Clare Borehole which helped to prove the depth of this buried valley, also gave the first firm evidence for the nature of the 'basement rocks' which underlie the Cretaceous and Tertiary strata in the district. The 'basement' at Clare forms part of the London Platform and consists of indurated Silurian mudstones, siltstones and sandstones which may constitute a future resource of hard-rock aggregate for south-east England. An outline of the major structures in the platform of the region, which includes Devonian as well as Silurian rocks, is being built up using gravity anomaly, aeromagnetic and other data; this will be vital information for any future exploitation of hard-rock resources in this area.

Peter J Cook, DSc
Director

Kingsley Dunham Centre
British Geological Survey
Keyworth
Nottingham
NG12 5GG

18 June 1993

HISTORY OF THE SURVEY OF THE SUDBURY SHEET

The district was originally geologically surveyed on a scale of one inch to one mile (1:63 360) by F J Bennett, J H Blake, W H Dalton, W H Penning and W Whitaker, and the results published on the Old Series maps 47 (in 1884), 48 (1882), 50 (1881) and 51 (1882). Explanatory memoirs covering the parts of those maps relevant to this district were published in 1878, 1885, 1881 and 1886 respectively.

The area around the town of Sudbury was resurveyed at a scale of six inch to one mile (1:10 560) by Professor P G H Boswell in 1911–13. The results were incorporated in the New Series geological map at a scale of one inch to one mile, covering the district described in this memoir and published in 1928. Professor Boswell wrote a memoir to accompany it (1929); a revised edition of the map was issued in 1947.

The resurvey of the whole district at 1:10 560 or 1:10 000 scale, upon which this memoir is based, began in 1974 as overlap from the resurveys of the adjoining Braintree (223) and Bury St Edmunds (189) sheets, to the south and north respectively. It continued in 1978–79 as a foundation for sand and gravel projects carried out by the Industrial Minerals Assessment Unit on behalf of the Department of the Environment. The remaining areas were resurveyed between 1980 and 1987. The resurvey was carried out by Dr N G Berridge, Dr C R Bristow, Mr R A Ellison, Mr M J Heath, Mr S R Mills, Dr D Millward, Dr B S P Moorlock, Mr J Pattison, Mr A Smith, Mr B Young and Dr J A Zalasiewicz under the supervision of Drs R A B Bazley, W A Read and R G Thurrell as successive Regional Geologists. The 1:50 000 Sudbury (206) Geological sheet was published in 1991.

The following is a list of the 1:10 000 and 1:10 560 geological maps included wholly or in part within the district, with the initials of the surveyors and the date of survey.

TL63NE	Cornish Hall End	JAZ	1980
TL64NE	Haverhill	NGB	1987
TL64SE	Steeple Bumpstead	NGB	1987
TL65NE	Kirtling	AS	1987
TL65SE	Great Bradley and Great Thurlow	NGB	1987
TL73NW	Stambourne	SRM	1974
TL73NE	Castle Hedingham	MJH	1975
TL74NW	Hundon and Kedington	DM	1978–79
TL74NE	Clare	DM	1978–79
TL74SW	Wixoe and Stoke by Clare	DM	1978
TL74SE	Belchamp St Paul	DM	1978
TL75NW	Lidgate	BSPM	1977
TL75NE	Chedburgh	BSPM	1976
TL75SW	Cowlinge and Stradishall	NGB	1987
TL75SE	Denston	JP	1987
TL83NW	Gestingthorpe	MJH	1974
TL83NE	Lamarsh	SRM	1974
TL84NW	Cavendish and Glemsford	RAE	1978
TL84NE	Long Melford	DM	1978
TL84SW	Foxearth and Belchamp Walter	RAE	1978
TL84SE	Sudbury	DM	1978
TL85NW	Rede and Whepstead	CRB	1976
TL85NE	Bradfield Combust	CRB	1976
TL85SW	Hartest	JP	1986–87
		BY	1980
TL85SE	Shimpling	BY	1980
TL93NW	Assington	SRM	1975
TL93NE	Stoke by Nayland	RAE and	
		SRM	1974–77
TL94NW	Lavenham	NGB	1986
TL94NE	Monks Eleigh	AS	1986
TL94SW	Great Waldingfield	NGB	1986
TL94SE	Boxford	AS	1986
TL95NW	Felsham	CRB	1976
TL95NE	Rattlesden	CRB	1976
TL95SW	Cockfield and Preston	BY	1980
TL95SE	Brettenham and Kettlebaston	CRB and	
		DM	1979
		AS	1986

ACKNOWLEDGEMENTS

Most of the memoir has been written by Mr J Pattison and Dr N G Berridge, with major contributions by Mrs J M Allsop on structure and pre-Cretaceous geology and Dr I P Wilkinson on Cretaceous stratigraphy. Mr Pattison also compiled the memoir. Other contributors include Dr D Millward and Messrs R A Ellison and A Smith on local stratigraphy, Mr R J Merriman on the petrology of some of the Palaeogene rocks and Mr A A Morter on Gault and Cambridge Greensand biostratigraphy. For the economic geology chapter, the sections on sand and gravel resources and hydrogeology were written by Mr P M Hopson and Mr N S Robins respectively. We are indebted to Mr C J Wood of Scops Geological Services Ltd for identifying Chalk macrofossils and for his invaluable advice on Upper Cretaceous stratigraphy. Thanks are also due, for their advice and contributions, to the following: (on stratigraphy) Dr C R Bristow, Mr M J Heath, Dr R W O'B Knox and Mr B Young; (on biostratigraphy) Mr F G Berry, Mr D K Graham, Dr R Harland, Mr M J Hughes, Dr M P Kerney, Dr A W Medd and Dr S G Molyneux; (on petrology) Mr R K Harrison and Mr A C Morton. The memoir has been edited by Dr C R Bristow.

Anglian Water Services Ltd and the National Rivers Authority (Anglian Region) provided the borehole resistivity logs and we are grateful for their permission to publish some of them in this memoir. We would also like to thank all the other organisations who provided borehole logs, and local landowners for their co-operation during the resurvey.

NOTES

The word 'district' is used in this memoir to denote the area included in the 1:50 000 Geological Sheet 206 (Sudbury).

National Grid references are given in square brackets throughout the memoir. Unless otherwise stated, all lie within the 100 km square TL.

The authorship of fossil species is given in the fossil index.

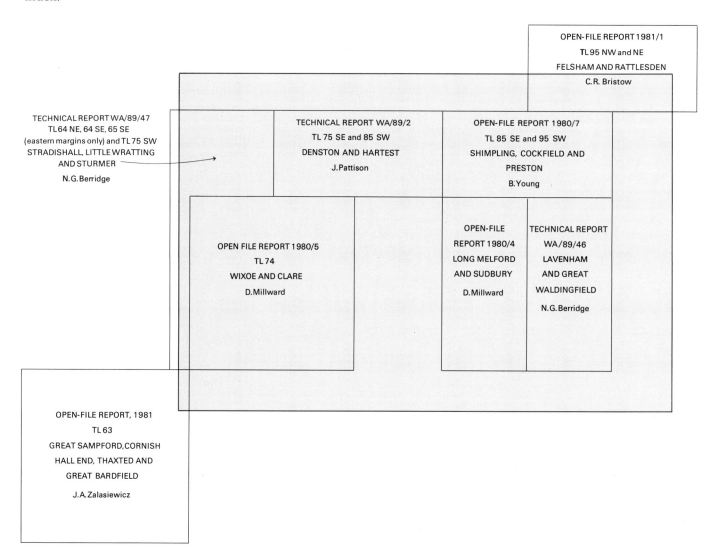

Plans showing geographical cover of BGS Onshore Geological Technical Reports (WA Series) and Open-File Reports in the district.

ONE

Introduction

This district lies mostly in Suffolk but includes part of Essex, south of the River Stour (Figure 1). The greater part of it is a low plateau formed by glacial till, commonly known as 'Chalky Boulder Clay'. The plateau rises from about 60 m above sea level in the south-east corner of the district to over 120 m in the north-west and includes some of the highest parts of East Anglia (Plate 1).

The Drift deposits below the till and the solid rocks underlying them commonly crop out along the lower flanks of the deeper valleys. These older Drift deposits mostly consist of sand and gravel, either of glacial origin or, where they form the Kesgrave Sands and Gravels, thought to have been deposited in the braided channels of a periglacial proto-Thames river. Above a basement of Silurian and Devonian strata, the solid rocks of the district include the Chalk, the Thanet Beds and Woolwich and Reading Beds (together referred to as the 'Lower London Tertiaries' where they cannot be separated), the London Clay and the Crag, in ascending sequence (Figure 2). The Chalk underlies the whole district, the 'Lower London Tertiaries' and the London Clay overlie the Chalk to the south and east of Sudbury, and the marine sands and gravels of the Crag have a patchy and somewhat indeterminate distribution below the Drift, mostly across the central and eastern areas. The only minerals of any current economic significance are sands and gravels belonging to

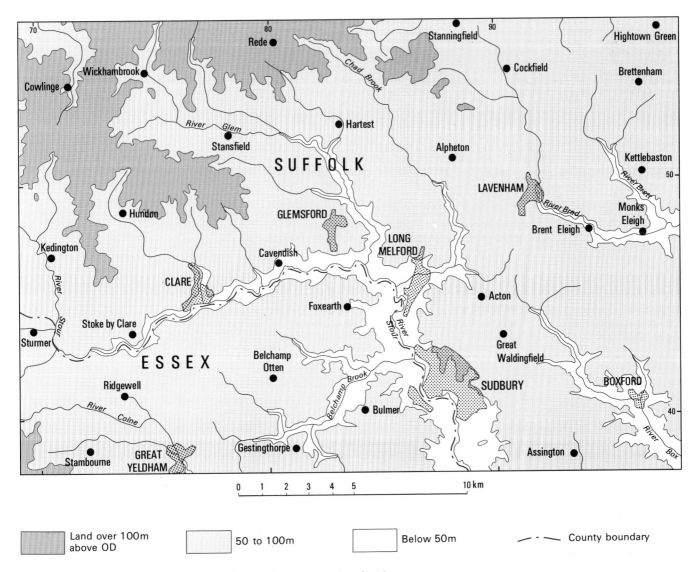

Figure 1 Sketch map showing the physical features of the district.

Plate 1 A valley cut into the west Suffolk till plateau at Hartest [8372 5220].

The ploughed field in the foreground shows a typical 'chalky till' brown clay soil with abundant flints and chalk pebbles. The highest point in Suffolk, at about 128 m above OD, appears on the horizon about 3 cm from the left hand edge. [A14833]

both Drift and Crag. The Chalk was previously extensively quarried, largely for agricultural lime, in the south-eastern parts of the district, especially around Sudbury town. It is the principal aquifer of the district.

Most of the district is drained by the River Stour and its tributaries, the River Glem, Chad Brook and Belchamp Brook, which all join it near Sudbury and Long Melford, and by the rivers Brett and Box, which flow through the eastern areas to meet the Stour outside the district. The headwaters of the River Colne flow across the south-western corner of the district, and some north-eastern areas are drained by tributaries of the River Gipping. The main East Anglian watershed curls across the higher parts of the plateau in the north-west, beyond which small areas are drained via the rivers Lark and Kennett towards the Great Ouse.

Arable farming is the principal land use and is the mainstay of the local economy. The commonest crops are wheat, barley, rape and sugar beet. Nevertheless, the rural landscape is pleasantly varied with water meadows along the main streams, scattered woodland for game conservation, and minor crops such as apples; willows are locally cultivated for making cricket bats. Traditional industries like woollens at Sudbury and linen at Glemsford have largely been superseded by light engineering, principally at Sudbury, and food processing, notably at Kedington, near Haverhill. Haverhill town, which is just beyond the western edge of the district, and Sudbury are both designated expanding towns, having taken London overspill population after World War II. Sudbury is the present-day terminus of a commuter railway to Colchester and London, as well as the former head of navigation on the River Stour. As an ancient town and the birthplace of the artist Gainsborough, it receives an increasing number of tourists, as do the smaller towns of Clare, Lavenham and Long Melford. The visitors are attracted by the fine flint-built churches (Plate 9) and half-timbered houses dating from the heyday of the late

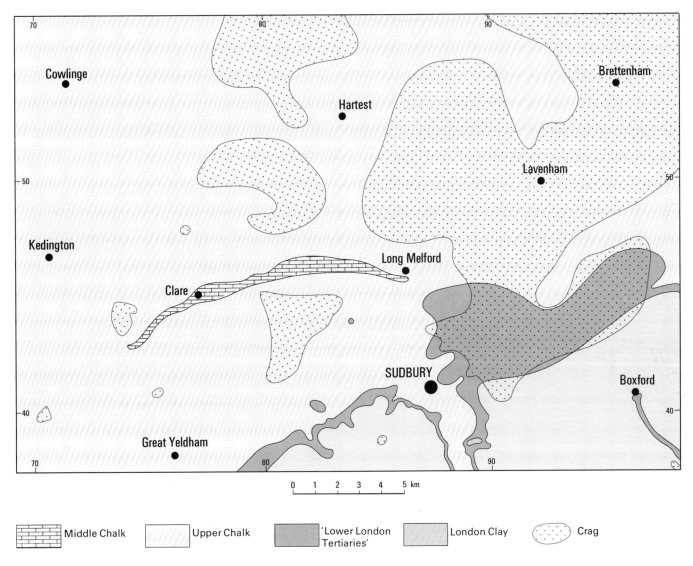

Figure 2   Simplified map of the solid geology.

medieval and Tudor woollen industry, and by the considerable number of antique shops.

HISTORY OF RESEARCH

The extensive drift cover has inhibited study of the local solid geology and, although exposures of glacial deposits near Sudbury were formerly described as the finest of inland drift sections (Whitaker et al., 1878), the district has been only peripheral to most Quaternary research in East Anglia. An outline of the geology was first shown in William Smith's maps of Suffolk (1819) and Essex (1820). The latter included a note on the succession of strata at Sudbury but, until the primary survey of the 1870s and 1880s, studies relevant to the district were generally confined to regional works such as those of Prestwich (1850, 1852, 1854) on the Palaeocene rocks and Searles Wood (junior) (1880) on the glacial deposits.

With help from the local amateurs, Dr Holden of Sudbury and the Reverend E Hill of Cockfield, Jukes-Browne (1903) classified the Chalk outcrops of Suffolk in terms of the nationally recognised macrofossil zonation and thereby demonstrated the scale of the Cretaceous/Tertiary unconformity. In several papers, Hill (1891, 1902, 1910, 1912) also catalogued data on exposures of till in the district. His 1912 paper included a particularly valuable list of the pits in and around Sudbury town. Since Jukes-Browne's pioneering work, the Chalk has received little attention until recent years, when correlation of borehole resistivity logs (Murray, 1986) and foraminiferal biostratigraphy (Bailey et al., 1983) have rendered further advances possible.

Work on the Palaeogene strata of Suffolk and north Essex has commonly been part of studies related to the whole of the London Basin. Prestwich, in a series of papers (1850, 1852, 1854), established the names of the main formations and discussed their biostratigraphy. A synthesis of the Survey's results on the London Clay and

'Lower London Tertiaries' was produced by Whitaker (1872), who himself supervised the primary survey of the district. Boswell (1916), in one of several papers covering much of southern East Anglian stratigraphy, showed the Palaeogene subcrop extending northwards below the Drift and Crag of the coastal areas. He also discussed the synsedimentary structures affecting the Tertiary rocks, a theme later taken up by Wooldridge (1923, 1926). The post-war years brought two review papers: one on the Woolwich and Reading Beds (Hester, 1965), the other on the Palaeogene of south-east England (Curry, 1965). More specialist papers on the Palaeogene strata in recent years include studies of sandstone provenance and diagenesis (Morton, 1982), facies variations in the Woolwich and Reading Beds (Ellison, 1983) and London Clay litho- and biostratigraphy (King, 1970, 1981; Knox et al., 1983).

Although the Pliocene and Pleistocene Crag marine deposits and fossils of East Anglia stimulated much study during the early and middle parts of the nineteenth century, none of it referred to this district because the south-westerly extension of the Crag outcrop into west Suffolk and north-west Essex was not recognised until the primary survey was under way. The Survey's findings on the Crag were summarised by Reid (1890), who included details and faunas from local sections. Harmer (1900, 1902, 1910) wrote on the Pliocene and Pleistocene stratigraphy of East Anglia and established molluscan-based biozones and stages founded on the ideas of progressive climatic cooling and a northward-younging of the Red and Norwich Crag outcrops. An assignment of those formations to the Quaternary (Pleistocene) rather than the Pliocene was proposed by Boswell (1952). Harmer's stratigraphical scheme survived until the drilling of a deep borehole at Ludham, Norfolk, through a thick Lower Pleistocene sequence. Pollen and foraminifera from this borehole, together with material from other East Anglian localities, were used to erect a new chronostratigraphical classification (West, 1961; Funnell, 1961; Funnell and West, 1977). The problems of mapping Crag have remained, however, as illustrated by the divergent conclusions of two recent papers relating to its distribution in this and surrounding districts (Bristow, 1983; Mathers and Zalasiewicz, 1988).

One of the main difficulties of mapping Crag in the inland areas of southern East Anglia has been distinguishing it from overlying early to middle Pleistocene glacigenic and periglacial fluviatile sands and gravels. The latter were all consigned to 'Glacial Sand and Gravel' in the primary survey of the district. From the late nineteenth century onwards, however, various post-Crag/pre-Lowestoft Till sand and gravel units have been described and named in East Anglia, including the Westleton Beds (Prestwich, 1890), Danbury Gravels (Gregory, 1915), White Ballast or Essex White Ballast (of authors including Mitchell et al., 1973) and the Westland Green Gravels (Warren, 1957) although most of these names are no longer used. The most recent Survey work in south Suffolk and north Essex has followed the nomenclature of Rose et al. (1976) who recognised

proto-Thames river sands and gravels of Beestonian age (Kesgrave Sands and Gravels), overlain by fluvioglacial gravels of Anglian age (Barham Sands and Gravels), although the latter are of limited distribution and commonly indistinguishable from slightly younger glacial deposits.

Much research on the widespread East Anglian tills has sought to discover the number of ice advances involved. In the 1860s, a number of short papers and letters, principally to the Geological Magazine, by Searles Wood (junior), Maw and Harmer concerned the interpretation of the East Anglian coastal sections. Wood (1880), and later Harmer (1910), summarised their views on the areal and stratigraphical distribution of till in the region. These works divided the till into a 'North Sea Drift' confined to north Norfolk, and a younger, geographically widespread 'Boulder Clay' described as either 'Chalky' or 'Chalky-Kimmeridgic'. The provenance of the tills and the direction of the ice flow from which they were deposited were later interpreted using heavy-mineral assemblages (Solomon, 1932), the lithology of erratics (Baden-Powell, 1948) and their orientation (West and Donner, 1956). The consensus view in the early post-war years, that the greater part of East Anglia had been covered by ice sheets during both the Anglian and Wolstonian stages, was summarised by Mitchell et al. (1973), but this view was disputed in an addendum to that report by Bristow and Cox. The arguments for only a single widespread glaciation prior to the Devensian have been presented in a series of papers (Cox and Nickless, 1972; Bristow and Cox, 1973; Perrin et al., 1973, 1979).

Although artefacts and other relics of early man have been reported from a number of places in the district (Wymer, 1985), the only local site of major archaeological significance is at Brundon, near Sudbury (Moir and Hopwood, 1939).

OUTLINE OF GEOLOGICAL HISTORY

The oldest known rocks in the district are the alternating sandstones, siltstones and mudstones of Silurian age proved in the Clare Borehole. They were deposited on a stable 'Midland Platform' south-east of the Iapetus Ocean. Boreholes in adjacent districts have proved Devonian sandstones, with interbedded argillaceous beds, below the sub-Cretaceous unconformity. These boreholes, together with aeromagnetic data, suggest a north-west to south-east structural grain in the Silurian and Devonian strata, which is attributed to the Caledonian orogeny.

The district reverted to stability in the late Palaeozoic and, as part of the 'London Platform', probably remained above sea level throughout the Permian, Triassic and Jurassic periods. The mostly shallow, epicontinental seas which covered much of the English Midlands during the Jurassic progressively transgressed onto the London Platform. This gradual submergence was interrupted around the end of the Jurassic by a combination of tectonic uplift and a eustatic fall in sea level, as a result of which any later Palaeozoic or early Mesozoic strata which may have been deposited would have been largely removed by erosion.

The progressive inundation of the platform from the north-west recommenced during early Cretaceous times and reached the district during the Albian. It resulted in an incomplete Gault clay sequence (with most of the early and middle Albian unrepresented), which was laid down in a quiet, shallow sea fairly distant from the shore. The influence of the London Platform on deposition waned during the later Albian and early Cenomanian, and by the end of the Cenomanian much of southern Britain was flooded by the Chalk sea. This sea is thought to have been of uniformly moderate depth, between 100 and 600 m, but intervals of reduced sedimentation or even submarine erosion are indicated by indurated beds and chalk conglomerates.

The close of the Cretaceous was marked by uplift and gentle folding, which resulted in the removal by erosion of up to 300 m of Upper Chalk, representing six macrofossil zones in parts of the region. Then, for the first time, the North Sea Basin became a dominant influence locally and, after an emergence lasting up to twenty million years, the district became submerged beneath the sea. The oldest sediments deposited in it were the shallow-water, bioturbated sands of the Thanet Beds, overlain, after a hiatus, by the interbedded littoral sands and mottled, probably brackish-water clays, of the Woolwich and Reading Beds. The mottling suggests a fluctuating water table possibly linked to intermittent emergence above sea level. Another transgression brought deposition of the London Clay with a large cool-water benthonic marine fauna and, in contrast, numerous washed-in remains of a rich tropical land flora.

Further earth movements, peaking during the Miocene, produced an overall south-easterly tilt and possibly some minor east–west or ENE–WSW flexuring. This led to partial removal and planation of the Palaeogene strata and further erosion of the Chalk, although the former extent of the Palaeogene rocks beyond the present outcrop can only be locally inferred (Bristow, 1983).

After another emergence lasting at least 30 million years, during which the modern drainage began to evolve, the North Sea advanced westwards again and deposited the various Crag formations of East Anglia. The first deposits were cross-bedded shingle beds followed by finer-grained intertidal sands. Both occur partly in elongate troughs, which may be of either tectonic or erosional origin. The shelly Crag faunas indicate an initial cool climate which became progressively colder, a precursor of the Pleistocene ice age. These were the last marine sediments deposited in the district.

About one million years elapsed before further deposition took place during the Beestonian cold stage, when braided rivers flowing west to east or south-west to north-east across the region deposited the terrace-like spreads of the Kesgrave Sands and Gravels and associated formations. A locally reddened palaeosol has been recognised at the top of the Kesgrave Sands and Gravels; it possibly indicates a significantly warmer Cromerian interglacial stage. The palaeosol is covered by till or sand and gravel, the lowest deposits of the Anglian ice sheet, commonly with an erosive base.

The Anglian ice entered the district from the north and west, and advanced to a position 15 to 40 km beyond it. In addition to the great thickness of chalky till deposited by the ice, the associated meltwater deposited silt, sand and gravel on, within and below the ice sheet and gouged out deep subglacial troughs mostly along the lines of the pre-existing valleys.

After retreat of the ice, subaerial drainage resumed, diverted in places from the preglacial pattern. Any irregularities which may have been present on the till surface were removed by erosion to produce a monotonous plateau, although continued downwarping of the North Sea Basin has tilted this surface towards the east. After the profound effects of the Anglian glaciation, those of the later cold stages were less dramatic. The associated falls in sea level rejuvenated the drainage system at least twice, and periglacial solifluxion refilled the reincised valleys with unconsolidated head deposits, probably augmented by fine-grained windblown sand. River terrace deposits, possibly associated with both glacial and interglacial stages, were laid down by most of the major streams; some contain relics of early man. Since the last (late Devensian) cold stage, the rise in sea level has led to increased alluviation along the River Stour and its main tributaries.

TWO

Pre-Cretaceous rocks

The district forms part of the London Platform where a relatively thin sequence of Cretaceous and Tertiary strata rests directly on Palaeozoic rocks. Only one borehole within the district, at Clare, has penetrated the cover, although evidence on the nature of the underlying rocks is forthcoming from boreholes in adjacent districts and from geophysical data (see Figures 3 and 4). In its eastern part, which includes this district, the London Platform basement is thought to consist of basinal Lower Palaeozoic and Devonian rocks deformed by the Caledonian orogeny. (Allsop and Smith, 1988). However, west of a north-west to south-east line beneath western Essex, located approximately between the Ware [TL 3531 1398] and Cliffe Marshes [TQ 7185 7858] boreholes, the platform 'basement' rocks are inferred to be relatively undeformed Lower Palaeozoic strata of the Midlands Microcraton.

PRE-SILURIAN

Silurian and younger rocks in the region (as shown in Figures 3, 4, 7 and 17) are low in magnetite content; thus the magnetic basement, indicated at depths of more than 4 km using the method of Vacquier et al. (1957), is taken to be of older Palaeozoic or Precambrian rocks. The broad, low-amplitude aeromagnetic anomalies recorded in this region (Figure 4), which are also characteristic of East Anglia as a whole, may therefore have their origin in the variable distribution of magnetic igneous material within that basement.

SILURIAN

Silurian strata were proved in the Clare Borehole [TL 7834 4536] between the basal Cretaceous unconformity, at a depth of 232.3 m, and the bottom of the borehole. They comprise 32 m of pale to dark grey, laminated, rhythmically alternating sandstones, siltstones and mudstones, with an irregular weathered upper surface. Many of the beds are graded, and shelly layers containing horny brachiopods, gastropods, bivalves, orthocone nautiloids and ostracods form the base of some of the sandstone beds. The rocks are not cleaved, but bedding dips vary between 35° and vertical; the steeper figures are usually associated with slumps or loading structures.

Three samples, from 236.2 m, 241.66 m and 247.14 m depths, were collected for micropalaeontological analysis; they yielded miospores, acritarchs and chitinozoa. Dr S G Molyneux comments that the miospore species *Ambitisporites* cf. *dilutus* (from all three samples), *Apiculiretusispora* cf. *synorea* (from 247.14 m), *Archaeozonotriletes*? sp. (from 236.2 m) and *Retusotriletes* cf. *warringtoni* (from the top and bottom samples), plus the acritarch *Leoniella carminae*

(from 236.2 m), indicate a late Silurian (Ludfordian or Přídolí) age, certainly for the top sample and probably for the other two as well. Given the absence of diagnostic Přídolí miospore taxa, a Ludfordian age is more probable, although the microflora may represent an impoverished Přídolí assemblage. The postulated late Silurian age is approximately eqivalent to that suggested for beds below the Cretaceous unconformity in the Stowlangtoft Borehole [TL 9475 6882] in the Bury St Edmunds district to the north (Bristow, 1990).

Geophysical logs of the Clare Borehole have provided physical properties data (illustrated in Table 1 and Figures 5 and 6) for the Silurian rocks, and are used to interpret the surface and airborne geophysical surveys in the region. The natural radioactivity and high velocity (corresponding to high density) of the Silurian rocks are illustrated in Figure 5, which is a crossplot of sonic log values of interval transit time (the inverse of the velocity in microseconds per foot) against the gamma-ray log values. The histograms in Figure 6 illustrate the variations about the standard mean of the measured physical properties (sonic velocity, resistivity, density and gamma-ray emission) for the Silurian rocks in the borehole.

Silurian strata occur throughout the region but there is no firm evidence for the presence of Lower Palaeozoic rocks of other age. Structures with a generally NNE–SSW trend, inferred from the aeromagnetic data may affect the Silurian rocks (see Figure 17b). They include a possible anticline with its axis crossing the south-east corner of the district along which Wenlock strata may subcrop below the basal Cretaceous unconformity. Younger Silurian rocks, including those proved in the Clare and Stowlangtoft boreholes, subcrop to the west and east of the putative anticline. A parallel anticlinal structure may occur farther east, with subcropping Silurian strata of Wenlock/?Llandovery age which are proved in the Harwich [TM 2593 3278], Stutton [TM 1500 3340] and Weeley [TM 1474 2183] boreholes (Allsop and Smith, 1988). These boreholes are all east of the Ware–Cliffe Marshes line mentioned above and the Silurian rocks encountered in them are characteristic of the deformed basinal Lower Palaeozoic strata found east of the Midlands Microcraton.

The boreholes proving Silurian rocks below the basal Cretaceous unconformity in southern East Anglia are interspersed with others which prove Devonian strata, and a model of the sub-Mesozoic geology can be inferred using a combination of this borehole evidence with the gravity data (Figure 7). A positive area on the Bouguer gravity anomaly map north of the district can be related to the high density, near-surface Silurian sedimentary rocks under the Mesozoic cover, proved in the Stowlangtoft Borehole. A comparable positive area in the south-eastern parts of the Sudbury district also probably indi-

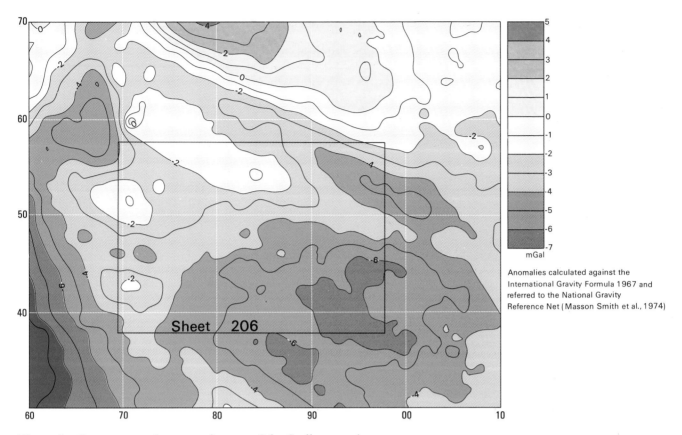

Figure 3   Bouguer gravity anomaly map of the Sudbury region.

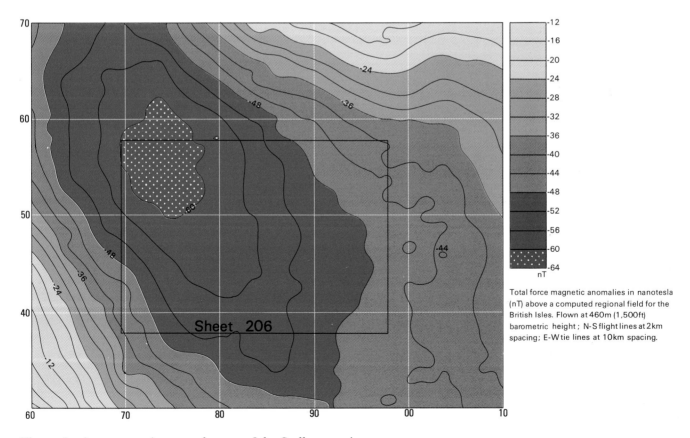

Figure 4   Aeromagnetic anomaly map of the Sudbury region.

Table 1 Physical properties of Mesozoic and basement rocks in southern East Anglia derived from analysis of the Clare Borehole geophysical logs and other sources†.

	Velocity km s^{-1}	Resistivity ohm m	Density Mg m^{-3}	Gamma Ray API units
CLARE BOREHOLE				
Middle Chalk (157–194)*	2.27 ± 0.33	1198 ± 62	2.22 ± 0.09	13.51 ± 6
Lower Chalk/Chalk Marl (194–222)	2.29 ± 0.17	1199 ± 72	2.32 ± 0.11	48.62 ± 21
Gault (222–232)	1.59 ± 0.11	300 ± 26	1.98 ± 0.20	111 ± 6
Silurian (232–263)	4.24 ± 0.29	1196 ± 70	2.71 ± 0.14	164 ± 15
AVERAGE FOR SOUTHERN EAST ANGLIA				
Silurian slates	5.01 ± 0.10	2000 ± 500	2.72 ± 0.10	200 ± 30
CANVEY ISLAND BOREHOLE				
Average—Devonian	–	–	2.49 ± 20	50 ± 8

* Depths below ground level in metres
† Allsop and Smith (1988)

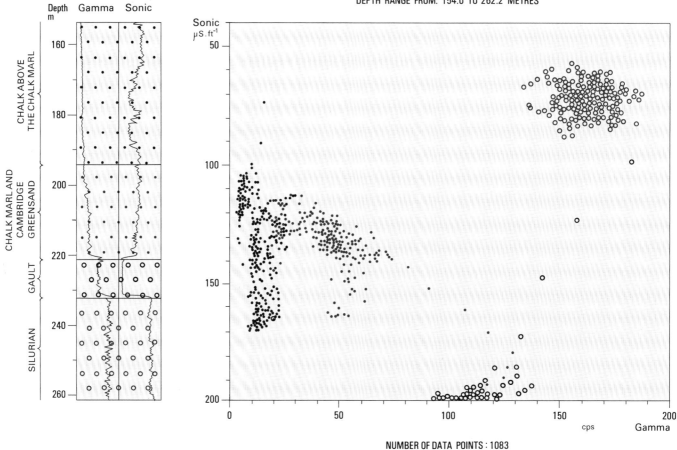

Figure 5 Gamma–sonic crossplot derived from downhole geophysical logs for the Clare Borehole.

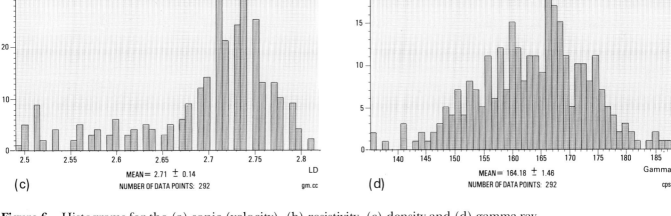

Figure 6 Histograms for the (a) sonic (velocity), (b) resistivity, (c) density and (d) gamma-ray logs for the Silurian rocks in the Clare Borehole.

cates the extent of Silurian rocks below the unconformity. The sub-Mesozoic surface is thought to dip gently south-eastwards at less than 1° (British Geological Survey, 1985; Bristow, 1990, fig.1).

Seismic refraction surveys in the Cambridge and Bury St Edmunds districts to the north also supplement the deep borehole data (Bullard et al., 1940; Bristow, 1990). The results from two sites in the Bury St Edmunds district were interpreted as top of basement refractors at depths of 0.21 and 0.16 km below OD, with seismic velocities of 4.9 and 5.1 km s[-1] respectively. These values, although very similar, are thought to represent contrasting sub-Mesozoic rocks, the first representing low density ?Devonian strata, the second indicating probable Silurian rocks.

DEVONIAN

Devonian strata have not been proved in the district but their presence is inferred from geophysical evidence. Negative areas on the Bouguer gravity anomaly map, in an east to west band across the northern half of the district and in the south-west corner, are taken to indicate troughs containing relatively low-density Devonian sedimentary rocks. The geophysical characteristics of the

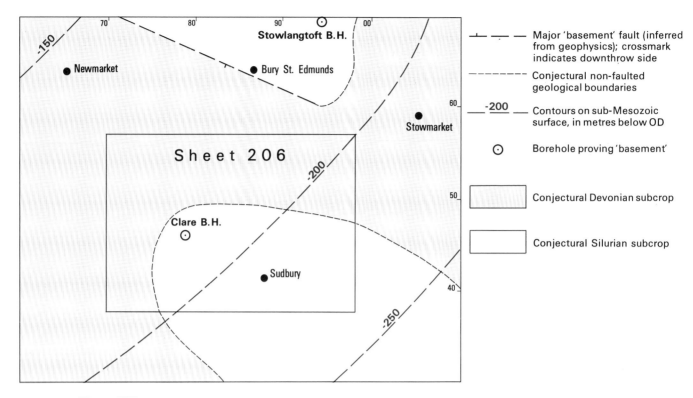

Figure 7   Sub-Mesozoic geology of the Sudbury region.

Devonian rocks in Southern East Anglia have been obtained from provings in several boreholes (Allsop, 1985), but notably at Canvey Island [TQ 8215 8330] (Smart et al., 1964). There, the interpolated downhole logs provided density values for the Devonian strata which vary widely, from 2.42 ± 0.02 Mg m^{-3} in the sandstones, to 2.59 ± 0.01 Mg m^{-3} in the shale horizons, giving a mean density for the complete Devonian sequence of 2.49 Mg m^{-3} (Allsop and Smith, 1988). Basement rocks of probable Devonian age, proved in the Lakenheath Borehole [TL 7480 8300] in the Thetford district and the Four Ashes Borehole [TM 0223 7186] in the Eye district, have density values of 2.62 ± 0.04 and 2.67 ± 0.03 Mg m^{-3} respectively (Chroston and Sola, 1982).

The seismic velocity of the Devonian rocks also varies with degree of compaction. Velocities of 3.5 to 4.4 km s^{-1} are typical but might not be applicable to the Sudbury district where lithologies of Lower Devonian and Upper Silurian strata may be similar.

Deep resistivity soundings in the region, made using the Wenner configuration with a maximum electrode separation of 1 km, included three sites [TL 562 586, 572 678, 583 629] just west of the Sudbury district. The results were interpreted using the method of Mooney and Wetzel (1957) and confirmed average resistivity values typical for Devonian rocks of 40 ohm.m and depths averaging 0.17 km below OD. This provides additional information on the possible extent and limits of the Devonian rocks in the western part of the district.

The geophysical characteristics of the Devonian rocks in the region are best characterised by those in the Can-

vey Island Borehole (Smart et al., 1964), which provides the best-documented long Devonian sequence in southern East Anglia, east of the Ware–Cliffe Marshes Borehole line mentioned above. It consists mainly of sandstones with interbedded siltstones, mudstones and conglomerates. Plant remains from several horizons in the borehole were dated as Downtonian (Přídolí) or Lower Devonian; Mortimer (1967) subsequently assigned the strata to the Emsian stage on spore evidence.

The (poorly developed) cleavage of the Devonian rocks in most of the borehole provings in this part of East Anglia suggest that a mild compressional phase occurred after deposition. Many of the steep 'dips' measured, for example 45° to 65° in the Soham Borehole [TL 5928 7448] (Worssam and Taylor, 1969), may represent cleavage rather than bedding planes.

TRIASSIC

The oldest Mesozoic rocks proved in the district belong to the Gault, which has a regional gentle south-south-easterly dip and which rests unconformably on the 'basement' rocks of the London Platform. However, the presence of small concealed pockets of low-density Triassic sediments below the Gault cannot be ruled out, because occurrences of red mudstones, sandstones and conglomerates of probable Triassic age are known to the north and west (Worssam and Taylor, 1969; Bristow et al., 1989; Bristow, 1990). Numerous minor closures on the Bouguer gravity anomaly map (Figure 3) as for example [850 420] west of Sudbury town, suggesting very localised and shallow low-density bodies, may indicate pockets of Triassic strata, but could also be due to variations in the thickness of Cretaceous units or small undulations of the basement surface.

THREE

Lower Cretaceous: Gault

The Gault is thought to underlie the whole district although it has been proved only in the Clare Borehole [TL 7834 4536] (Figure 8). The formation in East Anglia and south-east England consists of dark to pale grey mudstones and siltstones, rich in fossils. At some horizons (mainly in the Lower Gault and the basal part of the Upper Gault), shell beds and phosphatic nodule horizons occur, commonly associated with pronounced erosion surfaces and silty beds. These become less common higher in the Upper Gault, which was deposited in the deeper water of the Late Albian. Using a combination of lithological and macropalaeontological parameters, Gallois and Morter (1982) defined 18 (and locally 19) rhythmic units. Although difficulty is caused by the upper part of each rhythm being eroded to a lesser or greater extent by the succeeding one, particularly in the Lower Gault, correlation throughout the whole of eastern England is possible.

To the north and west of the Sudbury district, the stratigraphically lowest deposit to accumulate during the Albian transgression, the Carstone, is of *L. tardefurcata* and *D. mammillatum* zonal age (early Albian). This arenaceous deposit passes up into the argillaceous Gault which is of mid Albian age (*H. dentatus* Zone, *L. lyelli* Subzone) at its base. In the Clare Borehole, the Carstone is not present and the Gault is thin.

The proved thickness of the Gault to the west, at Duxford [TL 469 457], is 48.45 m, from where it thins towards the north-east to 18.92 m at Ely-Ouse No. 14 Borehole [TL 6962 8115] (Figure 9). However, a more rapid attenuation of the lower part of the formation takes place onto the London Platform, so that at Little Chishill [TL 4528 3637], where the Lower Gault may be missing, the formation is about 16.25 m thick and in the Clare Borehole [TL 7834 4536], where at least Beds 1 to 9 are missing, the Gault is only 11.08 m thick. The Upper Gault, on the other hand, shows little evidence of attenuation, and away from the thick sequence at Duxford (43.72 m thick), its thickness is remarkably consistent: 11.61 m, 12.63 m and 10.96 m in Ely-Ouse No. 14, Four Ashes (Morter in Bristow, 1980) and Clare boreholes respectively.

The basal 0.15 m of the Gault at Clare consists of a breccia comprising angular and weathered fragments of siltstone and black phosphatic nodules in a dark grey mudstone with *Chondrites*; it yielded no shelly macrofossils. This bed does not fit within the scheme erected by Gallois and Morter (1982), but the lithology of the mudstone matrix is similar to the topmost bed of the Lower Gault (Bed 10) elsewhere in East Anglia. Between 231.60 and 232.13 m, a bed of dark grey mudstone with *Chondrites* and phosphatic nodule horizons contains *Euhoplites* cf. *subcrenatus*, *Neohibolites minimus*, *Birostrina sulcata* and

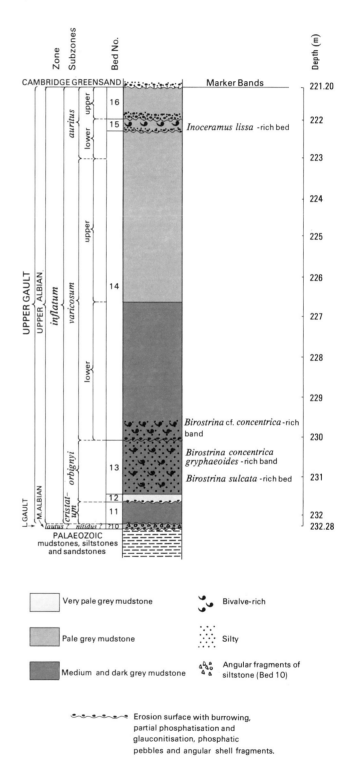

Figure 8 Stratigraphy of the Gault in the Clare Borehole.

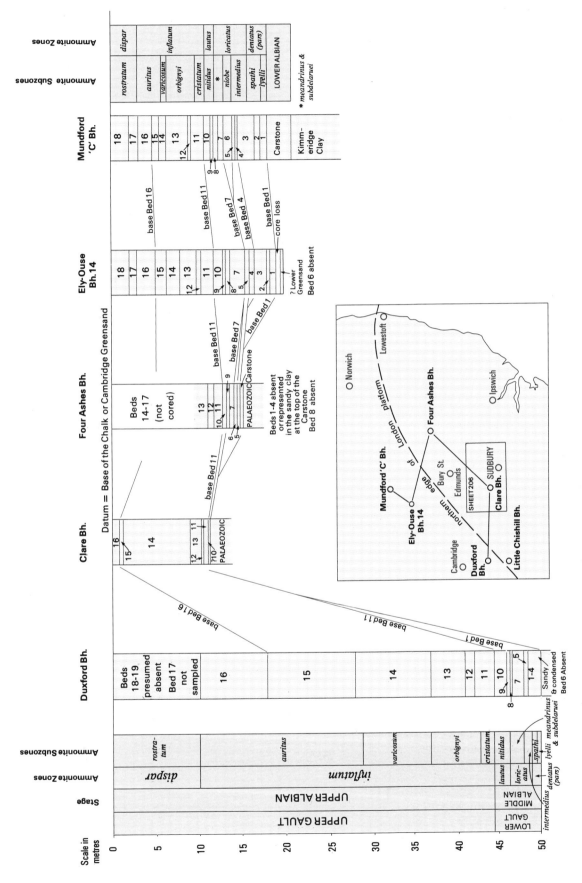

Figure 9 Correlation of Gault sequences proved in East Anglia.

serpulids. This fauna suggests Bed 11 (*M. inflatum* Zone, *D. cristatum* Subzone). Hence, the marine transgression that commenced in the early Albian elsewhere did not extend on to the London Platform in this area until mid to late Albian times.

The Upper Gault in the Clare Borehole compares closely with Beds 11 to 16 described by Gallois and Morter (1982), the *D. cristatum* to the *C. auritus* subzones being recognised. The top of the *C. auritus* Subzone and the whole of the *S. dispar* Zone were removed by erosion prior to the deposition of the Cambridge Greensand.

DETAILS

The following succession was recorded in the Clare Borehole (bed numbers are those of Gallois and Morter, 1982):

	Thickness m	Depth m
Bed 16 Pale grey calcareous mudstone with black phosphatic pebbles and an erosion surface at the base. Macrofauna includes *Aucellina coquandiana*, *Entolium orbiculare*, *Mortoniceras* (*Mortoniceras*) sp. and *?Holoaster* sp. Dr A W Medd reports the presence of the coccolith *Eiffellithus turrisseiffeli* and nannoconids. *Callihoplites auritus* Subzone (upper part).	0.78	221.98
Bed 15 Pale grey calcareous mudstone with abundant *Inoceramus lissa* and pale phosphates, becoming silty at the base. Basal part equivalent to the Barnwell Hard Band. Base marked by phosphatic nodules and an erosion surface. *Pycnodonte* (*Phygraea*) aff. *vesicularis*, *Mortoniceras* (*M.*) sp. and *Isocrinus legeri* are present. Dr A W Medd reports that the last occurrence of the coccolith proto-*Eiffellithus* is in this bed. *Callihoplites auritus* Subzone (lower part)	0.30	222.28
Bed 14 (*pars*) Pale grey calcareous mudstone with pale phosphate and inoceramid fragments. *Inoceramus lissa*, *Atreta* sp., *Pycnodonte* (*P.*) sp., *Anchura* (*Perissoptera*) cf. *maxima* and *Hysteroceras* sp. are present. *Callihoplites auritus* Subzone (lower part).	0.72	223.00
Bed 14 (*pars*) Lithology as above. Macrofauna includes *Birostrina* cf. *concentrica*, *Pycnodonte* (*P.*) aff. *vesicularis*, *Euhoplites alphalautus*, *Hysteroceras carinatum*, *H.* aff. *orbignyi*, *Mortoniceras* (*Deiradoceras*) cf. *devonse*, *Neohibolites minimus* and *N. oxycaudatus*.		

Inoceramid fragments common. (The smaller numbers of *Neohibolites* spp. suggests affinities with the Gault of southern England rather than that of East Anglia). *Hysteroceras varicosum* Subzone (upper part).

	Thickness m	Depth m
Hysteroceras varicosum Subzone (upper part).	3.52	226.52
Bed 14 (*pars*) Medium to dark grey mudstone with pale phosphates and dark grey phosphatic pebbles, becoming silty and shelly towards the base. Erosion surface with phosphatic nodules at the base. Macrofauna abundant and diverse, including: *Atreta* sp., *Cymbula?* cf. *phaseolina*, *Ludbrookia* sp., *Nucula* (*Pectinucula*) *pectinata*, *Pycnodonte* (*P.*) aff. *vesicularis*, common *Nerineopsis* cf. *coxi*, juvenile apporrhaids, *Anahoplites* sp., *Epihoplites* sp., *Euhoplites alphalautus*, *Hysteroceras* cf. *binum*, *H.* cf. *varicosum*, *Idiohamites* cf. *tuberculatus*, *Mortoniceras* cf. *pricei*, *Neohibolites minimus*, anthozoa, brachiopods and scaphopods *Birostrina* cf. *concentrica*-rich band at the base with common *B.* cf. *subsulcata*. *Hysteroceras varicosum* Subzone (lower part).	3.48	230.00
Bed 13 Medium grey silty mudstone with horizons of pale phosphates. Shelly *Birostrina sulcata*-rich bed, with *B. concentrica gryphaeoides* in the upper part and bands of *Inoceramus* cf. *anglicus* and *Neohibolites minimus* in the lower part. Ammonites include *Euhoplites inornatus*, *E.* aff. *trapezoidalis*, *Hysteroceras* cf. *orbignyi* and *Mortoniceras* (*D.*) cf. *albense*. *Hysteroceras orbignyi* Subzone.	1.37	231.37
Bed 12 Mottled pale and very pale grey mudstone, with a basal phosphatic nodule horizon resting on an erosion surface. Only *Nucula* (*P.*) *pectinata* recorded. Placed in the *Hysteroceras orbignyi* Subzone by comparison with other boreholes in East Anglia.	0.23	231.60
Bed 11 Dark grey mudstone with black phosphatic nodules at 231.96 m, just above a basal erosion surface. *Birostrina sulcata*, *Euhoplites* cf. *subcrenatus*, *Neohibolites minimus* and annelids are present. (The highest, *H. orbignyi* Subzone, part of the bed has been removed by erosion). *Dipoloceras cristatum* Subzone.	0.53	232.13
?Bed 10 Angular pebbles of siltstone and black phosphatic nodules in a pale grey matrix. Phosphatic nodule horizon rests unconformably on Lower Palaeozoic deposits. Macrofossils absent. *?Euhoplites nitidus* Subzone.	0.15	232.28

FOUR

Upper Cretaceous: Chalk (including Cambridge Greensand)

Chalk underlies the entire district and is at rockhead in at least half of it, but outcrops are restricted to a few narrow valley-side strips along the River Stour near Sudbury and in the valleys of the upper Stour, Belchamp Brook and the rivers Brad and Glem (Figure 10). Classification of the local succession is largely based on a combination of borehole resistivity log correlation and a small number of foraminiferal and macrofossil collections. A gentle east to south-east dip is indicated, averaging about 1 in 200, with a probable change of strike from north-east–south-west to about north–south near the middle Stour valley. A linear subcrop of Middle Chalk is inferred along the line of the Stour buried channel; Middle Chalk was proved there at rockhead in the Clare Borehole. The remainder of the subdrift crop is of Upper Chalk, which appears to range from the *Micraster cortestudinarium* to *Offaster pilula* macrofossil zones (Figure 11). The total Chalk thickness in the district is about 270 m. However, only for the lowest 27.34 m of the Lower Chalk, which was cored in the Clare Borehole, can a detailed lithological and biostratigraphical sequence be described.

The Chalk is generally pure, soft limestone largely composed of fragmental plates (coccoliths) of planktonic algae, with some beds of gritty texture consisting of comminuted inoceramid bivalve shells. It is remarkably free of terrigenous material, except the 'Chalk Marl' of the Lower Chalk. Some thin, relatively hard bands occur, such as the Totternhoe Stone, Melbourn Rock, Chalk Rock and Top Rock. They represent either hardgrounds, indicating pauses in sedimentation, or chalk conglomerates produced by submarine reworking.

Much of the Chalk seen in exposures consists of structureless, fine-grained chalk groundmass known as 'putty chalk', as for example at Ballingdon [860 405], where it is 20 m thick, and in the River Glem [8005 5207]. Up to 26 m of 'reconstituted' chalk were recorded near Stowlangtoft in the adjacent Bury St Edmunds district (Bristow, 1990, p.40). Pebble-grade hard clasts of apparently unaltered chalk may occur within this fine chalk matrix, but all evidence of bedding is commonly obliterated and the course of flint bands disrupted. Williams (1987) attributed 'putty chalk' to in-situ periglacial weathering, noting that on the Sussex coast it is largely confined to valleys, where the near-surface Chalk would have been affected by permafrost below a higher than present-day water table.

Large chalk rafts within till have been recorded in East Anglia, notably on the Norfolk coast (Peake and Hancock, 1961) and near Norwich (Wood, 1988), and the possibility that some mapped Chalk may merely represent large glacial erratics cannot be discounted. Chalk rafts at least 21 m thick have been proved in the Sudbury district by boreholes into the glacigenic fill of the Stour buried channel (Barker and Harker, 1984, pp.110, 112).

CAMBRIDGE GREENSAND

The Cambridge Greensand is traditionally considered to be the basal bed of the Chalk in Bedfordshire, Cambridgeshire, western Suffolk and Norfolk and is, therefore, regarded as part of the Lower Chalk. It comprises glauconitic sandy silts with phosphatic nodules at the base, but becomes less phosphatic upwards. It rests on the irregular eroded top of the Gault and is rarely more than about 0.3 to 0.6 m thick, although thicknesses up to approximately 1.5 m occur in the Cambridge district (Worssam and Taylor, 1969) and up to 1.8 m in the Bury St Edmunds district (Bristow, 1990). In the Sudbury district, the deposit has only been seen in the Clare Borehole in which it is 0.92 m thick.

The Cambridge Greensand contains a variety of pebbles, some purported to have originated in Wales (Hawkes, 1943), phosphatised moulds of shells, which have been mainly derived from the Gault, and an indigenous macrofauna. The reworking of fossils has created difficulties in dating the deposit. Penning and Jukes-Browne (1881), after mapping the Cambridge area, concluded that it was Cenomanian in age and they were supported by Reed (1897) and Peake and Hancock (1961). Spath (1923–1943), however, disagreed and, on the basis of the ammonites, considered the deposit to be of late Albian age, a view also held by Breistroffer (1940, 1947) and Casey (in Edmonds and Dinham, 1965). The foraminifera of the Cambridge Greensand at Barrington (Hart, 1973) suggest that it falls within the *Neostlingoceras carcitanense* ammonite Subzone (at the base of the *Mantelliceras mantelli* Zone). Wilkinson (1988) recorded the index species of the *Cythereis (Rehacythereis) luermannae hannoverana* ostracod Zone, *Planileberis scrobicularis* ostracod Subzone (Upper Albian), in the basal part of the deposit near Mildenhall, where the member is most complete. It was concluded that only the upper part of the Cambridge Greensand, in which several species of *Bythoceratina*, *Cytherella globosa* and *Cythereis (R.) bemerodensis* occur, should be considered of Cenomanian age.

The erosive base of the Cambridge Greensand rests variably on Beds 16 to 19 of the Gault (Morter and Wood, 1983). In the Clare Borehole (Figure 12), the Cambridge Greensand overlies Gault Bed 16 and comprises:

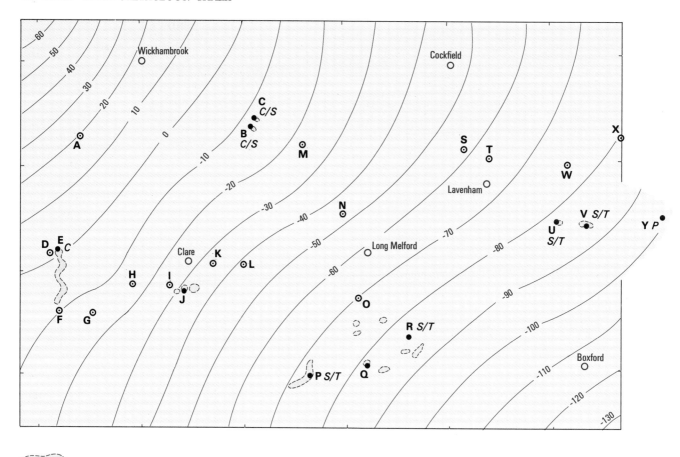

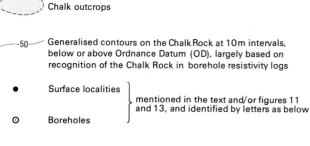

Chalk outcrops

—-50— Generalised contours on the Chalk Rock at 10m intervals, below or above Ordnance Datum (OD), largely based on recognition of the Chalk Rock in borehole resistivity logs

● Surface localities
⊙ Boreholes
} mentioned in the text and/or figures 11 and 13, and identified by letters as below

The zonal positions of some surface localities are indicated by the letters:

C   *Micraster coranguinum* Zone

C/S   *M. coranguinum* to *Uintacrinus socialis* zones

S/T   *U. socialis* to *Marsupites testudinarius* zones

P   *Offaster pilula* Zone

A   Highpoint, Stradishall Borehole [7189 5169]
B   River Glem [8005 5207]
C   Hawkedon Brook [8033 5240]
D   Kedington Borehole [7060 4592]
E   Old pit [7095 4612] near Baythorn Lodge
F   Wixoe Borehole [7093 4309]
G   Baythorne End Borehole [7247 4303]
H   Stoke by Clare Borehole [747 443]
I   Mill Farm Borehole [7621 4423]
J   South side of Stour valley [77 44] near Clare
K   Clare Borehole [7834 4536]
L   Bower Hall Borehole [796 453]
M   Boxted Borehole [8272 5119]

N   Cranmore Green Borehole [848 478]
O   Rodbridge Corner Borehole [8572 4398]
P   Old pit [8321 4006] near Goldingham Hall
Q   Old pit [860 405] at Ballingdon
R   Victoria Pit, Sudbury [8793 4175]
S   Alpheton Borehole [9042 5061]
T   Frog's Hall Borehole [9185 5017]
U   Old pit [9486 4714] near Wells Hall, Brent Eleigh
V   Old pit [9662 4717] at Swingleton Green, Monks Eleigh
W   Kettlebaston Borehole [9544 4996]
X   Hitcham Borehole [9829 5113]
Y   Old pit [988 475] near Nedging (on 1:50 000 sheet 207)

Figure 10   Chalk localities and structure contours on the Chalk Rock.

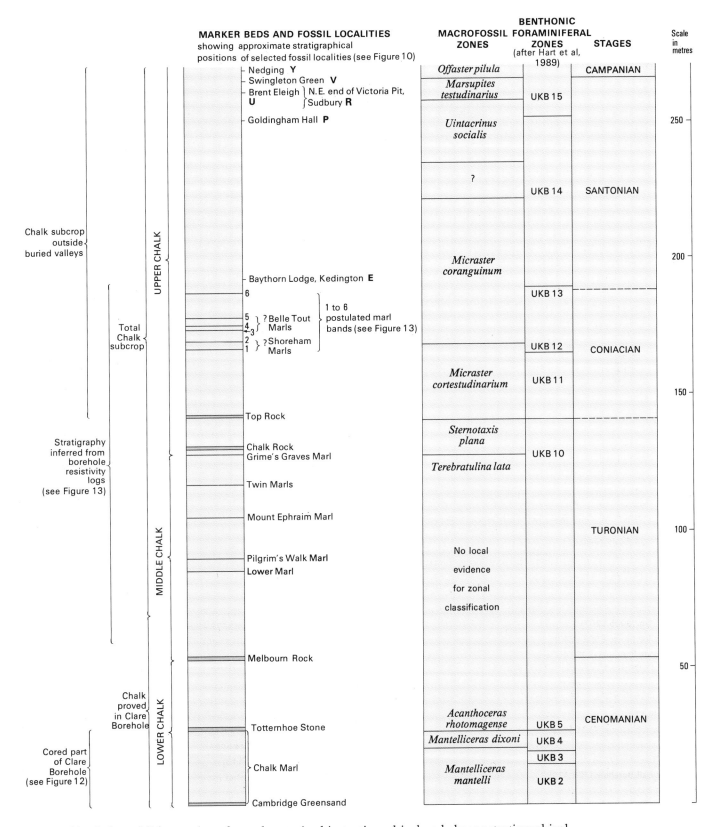

Figure 11 Inferred lithostratigraphy and tentative biostratigraphical and chronostratigraphical classification of the Chalk in the district.

	Thickness m	Depth m
Siltstone, pale grey, very calcareous, glauconitic, with sporadic black phosphatic nodules up to 5 mm across (size and frequency increases downhole); bioturbated, with glauconite-filled burrows; bivalves rare; gradational base	0.42	220.70
Siltstone, medium grey, calcareous; glauconitic (increasingly with depth); black phosphatic nodules; irregular cream phosphatic bodies up to 20 mm long; bioturbated, with glauconite-filled burrows and *Chondrites*; shell debris and bivalves, including *Aucellina*, present near the base; phosphatic nodules and erosion surface at base	0.17	220.87
Siltstone, pale grey, calcareous, glauconitic; phosphatic nodules up to 20 mm; bioturbated, with *Chondrites*; shelly, including common *Aucellina*, (notably *A. gryphaeoides*); concentration of phosphatic nodules at the base; basal erosion surface; burrowed into the top of the Gault	0.33	221.20

Foraminifera were collected from a sample between 220.50 and 221.06 m; long-ranging benthonic species such as *Tritaxia pyramidata*, *Gavelinella cenomanica*, *Gyroidinoides nitidus* and planktonic *Hedbergella delrioensis* dominate the assemblage. However, abundant specimens of *Arenobulimina anglica*, the index for the earliest Cenomanian foraminiferal zone of Robaszynski and Amédro (1980) are found associated with *Flourensina intermedia* and *Plectina mariae*, typical of the benthonic foraminifera Zone 8 (*sensu* Carter and Hart, 1977) and UKB2 (*sensu* Hart et al., 1989), approximately equivalent to the lower part of the *N. carcitanense* macrofossil Subzone. The presence of rare specimens of the ostracod species *Neocythere (Physocythere) steghausi* also suggest the lower part of that subzone.

CAMBRIDGE GREENSAND TO BASE OF MIDDLE CHALK

Lower Chalk strata above the Cambridge Greensand include the Chalk Marl and the Totternhoe Stone, both cored in the Clare Borehole, and a less well-known sequence between the Totternhoe Stone and the base of the Middle Chalk. The Clare Borehole proved 53.28 m of strata between the inferred base of the Middle Chalk and the top of the Cambridge Greensand. The lowest 25.91 m are assigned to the 'Chalk Marl' which has distinctive lithological and geophysical properties (Figure 5).

Porcellanous Beds

At the base of the Chalk Marl in the Clare Borehole are 13.78 m of pale grey, homogeneous, smooth-textured, sparsely fossiliferous, marly chalk, equivalent to the 'Porcellanous Beds' of Morter and Wood (1983) (Figure 12). *Aucellina* spp., comprising both the *gryphaeoides* and *uerpmanni* morphotypes recognised by those authors, range throughout the greater part of this sequence: the highest specimen recorded was at 209.18 m, 11.10 m above the top of the Cambridge Greensand and 0.23 m below the upper limit of beds with a noticeably porcellaneous texture. Apart from two questionable specimens from 208.60 m and 208.35 m, this horizon also marks the top of the range of '*Inoceramus*' ex gr. *anglicus–comancheanus*. The association of *Aucellina* and '*I.*' ex gr. *anglicus– comancheanus* indicates the lower part of the *Neostlingoceras carcitanense* Subzone of the *Mantelliceras mantelli* macrofossil zone. The remaining fauna is of relatively low diversity and includes *Kingena* cf. *arenosa*, *Monticlarella* sp., *Anomia* sp., *Entolium* sp. and the long-ranging *Plagiostoma globosum;* notable records are the occurrence of the giant isopod *Palaega carteri*, the 'shrimp' *Glyphaea willetti* and a superbly preserved juvenile *Hemiaster* ex gr. *griepenkerli–morrisi*, retaining the radioles. The 215 to 213 m interval is characterised by terebratulid brachiopods of uncertain affinities, including large forms at 214.34 m and 214.24 m. Foraminifera from the Porcellanous Beds in the borehole belong to Benthonic Formaminifera Zone 8 [Zone UKB2], recognised by the concurrent range of *Flourensina intermedia* and *Arenobulimina anglica*.

Chalk Marl above the Porcellanous Beds (including the First and Second Inoceramus Beds)

The Porcellanous Beds in the Clare Borehole (Figure 12) are succeeded by 12 m of off-white to pale grey marly chalk with common bivalve debris, divisible into eight or nine rhythmic units (Figure 12). Each unit begins at a well-defined, commonly burrowed erosion surface and is overlain by grey, gritty chalk, with fragments of white chalk in the basal few centimetres, grading up into less marly material to the top. The most conspicuous gritty beds, those of the lowest two units, are the two Inoceramus Beds of the condensed Cenomanian succession in eastern England (1 and 2 on Figure 12). The entry in flood abundance of '*Inoceramus*' *crippsi* at the base of the First Inoceramus Bed marks the base of the upper part of the *N. carcitanense* Subzone as interpreted by Gale and Friedrich (1989) in the Folkestone/Dover succession. The base of Benthonic Foraminifera Zone 9 [Zone UKB3], as recognised by the lowest occurrence of *Pseudotextulariella cretosa*, falls immediately beneath the base of the Second Inoceramus Bed in the borehole. However, it has been recorded by Dr K C Ball at the base of the First Inoceramus Bed at Hunstanton (personal communication).

The Second Inoceramus Bed, with its base at 204.57 m, consists of coarse, bioclastic chalk, containing

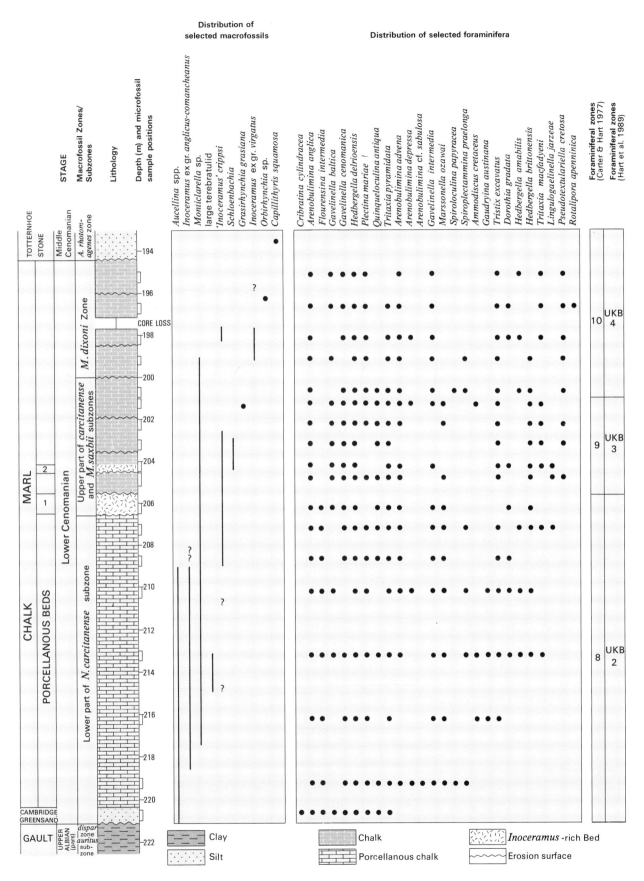

Figure 12   Biostratigraphical classification and distribution of selected macro- and microfossils in the Lower Chalk of the Clare Borehole. The numbers 1 and 2 indicate the First and Second Inoceramus beds.

lignite and inoceramid shell detritus. Poorly preserved *Schloenbachia varians* occur near the base of this bed and in the strata above it, up to 203.29 m depth, these being the only Upper Cretaceous ammonites noted in the borehole.

The lowest occurrence of *Inoceramus* ex gr. *virgatus* at 199.25 m, with five additional records between 198.25 and 197.65 m, suggests that the base of the *Mantelliceras dixoni* Zone could be drawn at the conspicuous burrowed erosion surface at 199.98 m. This surface would then equate with the terminal *Mantelliceras saxbii* Subzone erosion surface recognised in the Folkestone/Dover succession (erosion surface below marker horizon M4 in Gale, 1989, fig. 2), and the closely spaced records of *I.* ex gr. *virgatus* would represent the characteristic acme occurrence of that species group in the beds beneath the dixoni limestones (marker horizon M6 in Gale, 1989).

It is noteworthy that the base of Benthonic Foraminifera Zone 10 [Zone UKB4], as recognised by the highest occurrence of *Marssonella ozawai*, can be accurately placed in the Clare borehole at 201 m, i.e. 1 m beneath the presumed terminal *saxbii* Subzone erosion surface, and at a similar stratigraphical level to that shown by Hart et al. (1989). However, an unpublished reinterpretation of the biostratigraphy of the standard Aycliff Borehole at Dover by C J Wood suggests that this microfossil datum occurs higher, at the base of the acme occurrence of *Inoceramus* ex gr. *virgatus*, a little above the base of the *dixoni* Zone.

The occurrence of *Orbirhynchia* sp. at 196.23 m probably represents the lowest of the three *Orbirhynchia* acme occurrences recognised in southern England immediately above the dixoni limestones (see Gale, 1989, fig. 3) and well within the *Mantelliceras dixoni* Zone. This *Orbirhynchia* bioevent equates with the Lower Orbirhynchia Band (Jeans, 1968) of the condensed Cenomanian succession in eastern England.

No biostratigraphically significant macrofossils were recorded between the erosion surface at 195.96 m and the burrowed surface beneath the Totternhoe Stone at 194.37 m, but these beds are likely to be Lower rather than Middle Cenomanian in view of the absence of the distinctive Middle Cenomanian faunas.

Lower Chalk above the Chalk Marl

The uppermost 0.5 m of the cored sequence in the Clare Borehole consisted of hard, gritty chalk with scattered phosphatic pebbles resting on a burrowed surface of Chalk Marl. More abundant fragments of phosphatic pebbles were found above this level during percussion drilling, indicating that coring commenced in the basal part of the Totternhoe Stone. The cored part yielded a single specimen of *Capillithyris squamosa*, a brachiopod which in southern England is characteristic of and probably restricted to the Cast Bed in the lower part of the *Turrilites costatus* Subzone of the basal Middle Cenomanian *Acanthoceras rhotomagense* Zone (see Gale, 1989, fig. 3 and p.79). The equation of the Totternhoe Stone with the Cast Bed has been recognised at Hunstanton and elsewhere in eastern England. The top (uncored) 25 m of the Lower Chalk in the Clare Borehole was recorded as 15 m of alternating hard and soft chalk overlain by 10 m of hard chalk.

The basal 5.2 m in the Wixoe Borehole [7093 4309], which comprised firm, marly chalk, grey at the top, becoming paler below, is also assigned to the Lower Chalk.

MIDDLE CHALK

Most driller's logs merely indicate flinty white chalk of variable hardness throughout the Middle Chalk of the district. Therefore, the stratigraphy has been inferred from the resistivity logs of Anglian Water boreholes, mostly in the middle Stour valley area (Figure 13). It follows the work of Murray (1986) and is based largely on the recognition and correlation of low resistance 'spikes', interpreted as marl bed indicators, in the sixteen-inch normal resistivity trace. Bristow (1990) suggests that these laterally persistent beds may be volcanic fall-out deposits. However, research on laterally equivalent marl seams in southern England (Wray, 1991; Wray and Gale, in press) discounts a vulcanogenic origin in favour of enhanced sediment supply to the basin as a result of erosion following a temporary fall in sea level; this is an updated version of the model proposed by Jefferies (1963) for the Plenus Marls.

'Very hard chalk' recorded between 139.3 and 139.6 m in the Wixoe Borehole and 164.5 to 167.0 m in the Clare Borehole, is taken to be the Melbourn Rock. The resistivity logs for both the Wixoe Borehole, which went through the entire Middle Chalk succession (with a thickness of 73.5 m), and the Kedington Borehole [7060 4592], which proved most of it, are both somewhat smooth and uninformative, especially in the lower part of the subdivision. The Methwold Marl may be indicated by a low resistance feature at 134.5 m depth in the Wixoe log, and bands 'F$_2$' and 'G' of Murray (1986) by comparable lows in both the Wixoe and Kedington logs (Figure 13).

The top 40 m or so of the Middle Chalk, from the Pilgrim's Walk Marl to the Grime's Graves Marl, can be correlated on the resistivity logs with confidence along the Stour valley from the western edge of the district as far east as Glemsford. The thicknesses of the intervals between the Pilgrim's Walk Marl, Mount Ephraim Marl and the 'Twin Marls' are approximately constant in that area. However, the beds between the 'Twin Marls' and the Chalk Rock are more variable, both in total thickness (from 5 to 14 m) and inferred lithology. At Stowlangtoft [9475 6882], in the Bury St Edmunds district, this interval is 21 m thick (Bristow, 1990), suggesting that the equivalent beds in the Sudbury district represent a more or less condensed sequence. The impersistent marl bed near the top was referred to the Grime's Graves Marl by Murray (1986) who showed that the Sudbury district approximates to its southern limit, a conclusion supported by the local resistivity logs.

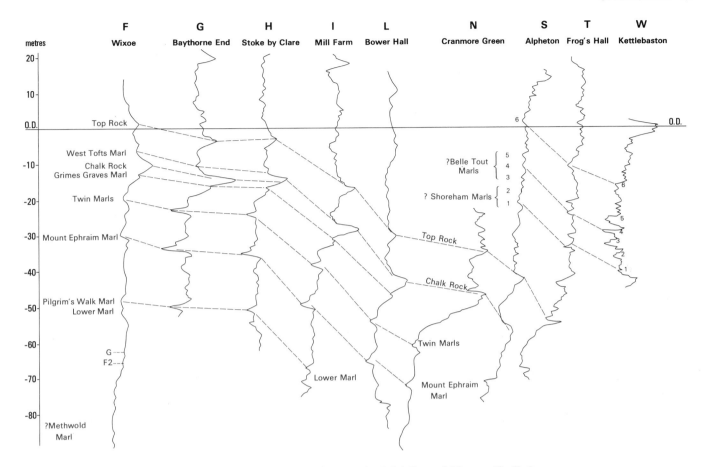

Figure 13   Correlation of 16-inch normal resistivity logs in the Middle and Upper Chalk for selected boreholes along the middle Stour valley and farther east. The borehole sites are shown on Figure 10. The method and nomenclature (except marl beds 1 to 6) follows that of Murray (1986).

UPPER CHALK

Because of the widespread drift cover and poor exposure, the local lithostratigraphy of the subdivision can only be inferred from the correlation of borehole resistivity logs and from a biostratigraphical and chronostratigraphical classification pieced together from the macrofossils and microfossils collected at a few scattered exposures. Consequently, for convenience, the following account is subdivided chronostratigraphically rather than lithostratigraphically (see Figure 11).

Turonian–Coniacian

In a regional context, flint content is greater in the Upper Chalk, notably in the basal few metres, than in the Middle Chalk, and this factor has been used locally to distinguish the boundary in some borehole lithological logs. Otherwise, the boundary can usually be fixed by recognition, in borehole resistivity logs, of the high resistance features made by the Chalk Rock and the slightly higher Top Rock. The lower bed is a little

above the base of the Upper Chalk (Figures 11 and 13). They may also be recognised, but less commonly, in lithological logs. Neither bed crops out in the district, but where they do a few kilometres to the north in the Bury St Edmunds district, the Chalk Rock is a thin sequence (less than 30 cm) of hard, yellow, nodular chalk, and the Top Rock comprises up to 40 cm of indurated, yellow, burrowed chalk with glauconite-coated chalk pebbles on its upper surface (Bristow, 1990). Both beds represent an attenuated succession, apparently relatively pronounced in the district. The combined thickness of chalk representing the *Sternotaxis plana* macrofossil zone, equivalent to the basal Upper Chalk up to the base of the Top Rock, and the overlying *Micraster cortestudinarium* Zone appears to be about 40 m, in contrast to 55 m in the Stowlangtoft Borehole. Nevertheless, the Chalk Rock/Top Rock interval remains fairly uniform at 11 to 13 m in all the Anglian Water boreholes from Wixoe to Cranmore Green [848 478], near Glemsford.

One possible consequence of attenuation in the lowest Upper Chalk may be the absence from much of the district of the Brandon Flint Series which underlies the

Chalk Rock in the Bury St Edmunds/Thetford area (Mortimore and Wood, 1986), although the 'hard and soft white chalk with a large amount of flint' recorded between 70 m and 90 m depth in a borehole [8572 4398] at Rodbridge Corner could indicate a local occurrence. The presence of the West Tofts Marl immediately overlying the Chalk Rock is suggested by most of the resistivity logs. In an uncommonly close correlation between lithological and geophysical logs, 'hard chalk' bands noted from depths 49.3 to 50.8 m and 70.2 to 71.4 m in the Boxted Borehole [8272 5119], which are taken to be the Top Rock and Chalk Rock respectively, approximate with high resistance features in the resistivity log.

The resistivity logs of several boreholes in the eastern part of the district show a number of low resistance 'spikes' which are taken to indicate marl beds at the top of the *Micraster cortestudinarium* Zone and in the lower (Coniacian) part of the *Micraster coranguinum* Zone; these are correlatable within and outside the district (Figure 13). Bands 1–2 and 3–5 probably equate with the Shoreham and Belle Tout marls respectively of the Anglo-Paris Basin (Mortimore, 1986). Based on these correlations, the *M. cortestudinarium–M. coranguinum* zone boundary, which in southern England is taken at the higher of the Shoreham Marls, is inferred to be at about 26 m above the Top Rock in boreholes at Alpheton [9042 5061] and Frog's Hall [9185 5017], near Lavenham.

Early Santonian

The stratigraphically lowest Chalk exposures in the district are in the Stour valley near Kedington and in the Glem valley near Hawkedon (Figure 10). The foraminiferal fauna in a sample from a 0.3 m section of chalk in an old pit [7095 4612] near Baythorn Lodge, Kedington, is poorly preserved and of low diversity, but includes *Lingulogavelinella arnagerensis* (= *L.* cf. *vombensis*), *Loxostomum eleyi*, *Neoflabellina suturalis*, ?*Reussella szajnochae praecursor*, *Stensioeina exsculpta exsculpta*, *S. exsculpta exsculpta–gracilis* (transitional), *S. granulata granulata* and *S. granulata polonica*. These species indicate the *Stensioeina granulata polonica* Foraminifera Zone (= lower part of UKB14 *sensu* Hart et al., 1989) and the presence of *L. arnagerensis* suggests that the sample was taken from the lower part of the zone. The Santonian part of the *M. coranguinum* macrofaunal zone may be inferred. In the resistivity log of the Kedington Borehole [7060 4592], about 430 m west-south-west of this locality, the Chalk Rock was identified at a depth of 60 m, thus providing an important link between resistivity and biostratigraphical evidence of the local Chalk stratigraphy, although the 61 m thickness of strata indicated between the Chalk Rock and the Santonian part of the *M. coranguinum* Zone is less than might be expected compared with the 85 m of similar strata at Stowlangtoft in the Bury St Edmunds district (Bristow, 1990).

Sections near stream level in Hawkedon Brook and the River Glem near Hawkedon consist entirely of 'putty chalk' with randomly scattered flints, and there must be some doubt whether the chalk seen there is in situ. Samples collected from two localities, at Hawkedon Brook [8033 5240] and River Glem [8005 5207], yielded foraminiferal assemblages of Santonian age. The Glem locality can probably be assigned to the highest part of the *M. coranguinum* zone or the basal *Uintacrinus socialis* zone. Similar doubt exists about Chalk mapped on the south side of the Stour valley near Clare, where the Chalk recorded both at surface and in boreholes may represent rafts within the till, a possibility supported by the relatively high rockhead indicated by those records.

The most conjectural interval of the Chalk succession shown in Figure 11 is the 50 m or so assigned to the poorly exposed middle and upper parts of the *M. coranguinum* Zone and the lower part of the *Uintacrinus socialis* Zone. In particular, the position of the *M. coranguinum/U. socialis* zonal boundary is uncertain. To the north, at Stowlangtoft, strata representing the *M. coranguinum* Zone are at least 55 m thick (Bristow, 1990). To the south, macrofaunal evidence indicates that the *U. socialis* Zone has a maximum thickness of 23 m in the Popsbridge Borehole [9706 3389], about 10 km southeast of Sudbury, within the Braintree district (Ellison and Lake, 1986). In the Sudbury district, a total thickness of just under 90 m for the combined zones is inferred from the regional dip. Any discrepancy between these figures might be explained by the northward thickening of strata in the '*Uintacrinus* Zone' in north Suffolk noted by Brydone (1932).

Late Santonian

A number of localities in the south-east quarter of the district have yielded broadly comparable foraminiferal assemblages, which are probably of late Santonian (mid *U. socialis* or *M. testudinarius* Zone) age. The localities include an old pit [8321 4006] near Goldingham Hall, Bulmer, with up to 3 m of rubbly chalk immediately overlain by Thanet Beds, and another [9486 4714], showing 0.5 m of Chalk, near Wells Hall, Brent Eleigh. The foraminiferal faunas at these two localities include *Cibicides beaumontianus*, *Gavelinella stelligera*, *G. cristata brotzeni* (= *G.* cf. *clementiana sensu* Bailey), *Globotruncana linneiana*, *Osangularia whitei*, *Reussella szajnochae praecursor*, *Stensioeina granulata incondita* and *S. exsculpta exsculpta*.

The former Victoria Pit, a large quarry [8793 4175] on the south side of the B115 road in Sudbury, is the best-preserved of these late Santonian localities. Up to 16 m of well-jointed, almost flintless chalk, much of it inaccessible, is exposed below Thanet Beds. Fossils were collected from the north-east end of the pit where 6.5 m of white, porcellanous, sparsely fossiliferous chalk is exposed. A sample 0.5 m above the base of the section yielded a moderate foraminiferal fauna including *Archaeoglobigerina cretacea*, *Gavelinella ammonoides*, *G. cristata brotzeni*, *G. stelligera*, *Stensioeina granulata incondita* and *S. granulata perfecta*. The last named is the index fossil for the eponymous foraminiferal zone equating with the middle part of the *U. socialis* macrofossil zone (Bailey et al., 1983). A conspicuous, non-indurated sponge bed about 2 m above the base contains hexactinellid sponges in ferruginous preservation, including *Lepi-*

dospongia? and *Rhizopoterion?*. The strata, including the sponge bed, between 1 and 2.5 m above the base of the section, also yielded *Terebratulina striatula, Mimachlamys cretosa, Pseudoperna boucheroni* (*sensu* Woods *non* Coquand), *Pycnodonte* sp., *Spondylus* cf. *latus* or *dutempleanus* and *Bourgueticrinus* columnals. Macrofossils collected from the top 2.5 m of the section include *Pycnodonte vesicularis, Pseudoperna boucheroni* and two fairly smooth *Marsupites* calyx plates, one of them slightly fluted. A sample 0.7 m from the top of the section yielded the following foraminifera: *Bolivinoides strigillatus, Cibicides beaumontianus, Globigerinelloides rowei, Reussella szajnochae praecursor* and *Globotruncana bulloides,* in addition to the foraminiferal species recorded from the lower sample. *B. strigillatus* is restricted to the *Marsupites testudinarius* and uppermost *U. socialis* macrofossil zones. *G. rowei* became extinct at the *M. testudinarius/O. pilula* boundary.

The combination of the foraminiferal and macrofaunal evidence therefore indicates that this 6.5 m section straddles the *Uintacrinus socialis–Marsupites testudinarius* zone boundary. The foraminifera indicate that it crosses the *polonica–strigillatus* boundary which is just below the *socialis–testudinarius* macrofossil boundary (Bailey et al., 1983). The *Marsupites* plates from the top part of the section confirm the *M. testudinarius* age, and the weak ornamentation suggests a low position in the zone. However, firm macrofossil evidence for assigning the lower part of the section to the *U. socialis* zone is lacking, although this pit may be that 'north-east of' Sudbury referred to by Jukes-Browne (1903, p.89) who recorded the belemnite *Actinocamax verus,* a species which is particularly common in, and characteristic of the zone, together with '*Lamna appendiculata, Oxyrhina* sp., *Actinocamax granulata, Ostrea semiplana* and *Lima hoperi*'.

Jukes-Browne (1903, p.89) also mentioned a chalk pit [9662 4717] at Swingleton Green, near Monk's Eleigh, from which a *Marsupites testudinarius* calyx plate had been collected. This quarry, later described by Boswell (1929, p.10), is now almost filled in. The remaining face shows 1.2 m of blocky, tough, apparently flintless chalk from which the following macrofossils were collected during the resurvey: *Porosphaera globularis, Gryphaeostrea canaliculata,* common thick-shelled inoceramid shell fragments including *Platyceramus?,* small *Bourgueticrinus* columnals, minute crinoid secundibrachial ossicles of either *Marsupites* or *Uintacrinus,* and a small shark's tooth crown (*?Scapanorhynchus subulatus*). The section also yielded a foraminiferal assemblage including *Stensioeina granulata perfecta,* which has been used as a middle *U. socialis* zone marker fossil,

and *S. exsculpta exsculpta* which, though ranging into the Campanian, is more characteristic of the Santonian.

The Hitcham Borehole [9829 5113], on the eastern edge of the district, penetrated 1.5 m of white, soft, shelly Chalk below the Crag. It yielded the foraminiferal species *Osangularia whitei, Reussella szajnochae praecursor, Rugoglobigerina* cf. *pilula, Stensioeina exsculpta exsculpta, S. granulata incondita* and *S. granulata perfecta,* which limit its possible age range to the middle to top of the *U. socialis* zone and the *M. testudinarius* zone.

Campanian

Jukes-Browne (1903) recorded a specimen of *Actinocamax granulatus* from near Nedging, just east of the district, and inferred the presence of the *Actinocamax quadratus* Zone in the sense then used, i.e. including both the *Offaster pilula* Zone and the restricted *Gonioteuthis quadrata* Zone of modern usage. The locality, a large and now mostly overgrown old chalk pit [988 475], was visited during the resurvey. Only three small exposures of soft white chalk with several oyster-rich beds were found. From these sections, which span an estimated stratigraphical thickness of 2.5 m, the following macrofauna was collected: *Ostrea incurva,* elongate forms of *Pseudoperna boucheroni, Spondylus* spp. ex gr. *latus* and *S. dutempleanus,* small inoceramid fragments and small *Bourgueticrinus* columnals. The elongate *P. boucheroni* suggest a late *M. testudinarius* or early *O. pilula* age, but the absence of *Marsupites* and the relative abundance of oysters favour the latter. This interpretation is supported by foraminiferal faunas which firmly date the Chalk here as early Campanian and, by inference, the *O. pilula* Zone. They include patchily distributed *Bolivinoides culverensis, B. strigillatus, Reussella kelleri, R. szajnochae praecursor, Stensioeina exsculpta exsculpta* and *S. granulata perfecta.* The two named species of *Bolivinoides* have a concurrent range in the *O. pilula* Zone (Bailey et al., 1983). Combining the microfossil and macrofossil evidence, it is possible that this locality falls in the oyster-rich beds in the lower half of the *Echinocorys depressula* Subzone of the *O. pilula* Zone (see Young and Lake, 1988, fig. 18). On this basis, it is likely that the belemnite recorded in 1903 was *Gonioteuthis granulata quadrata.*

The Campanian age indicated for the rocks in this quarry, and its location just outside the Sudbury district but along the inferred strike from places well inside it, suggest that Campanian strata may well subcrop below the Tertiary rocks in the south-east of the district.

FIVE

Palaeocene and Eocene: 'Lower London Tertiaries' and London Clay

The Palaeocene and Eocene strata, which together comprise the Palaeogene (Table 2), lie unconformably on the Chalk in the south-east quarter of the district. The truncated beds below the unconformity are of Upper Chalk age, ranging from the *Offaster pilula* Zone downwards, probably to and including much of the *Uintacrinus socialis* Zone. The plane of the unconformity mostly dips south-south-eastwards at about 1 in 200 (Figure 14). The Palaeogene strata are about 5 m thick or less around Acton and Milden, but from a west-south-west to east-north-east line through Sudbury they thicken south-south-eastwards to over 40 m south of Boxford (Figure 15). These thickness variations are probably, in the main, a result of post-Palaeogene erosion, but a south-south-easterly continuation of thickening in the 'Lower London Tertiaries' strata in the Braintree district to the south suggests that some of them may be depositional or intra-Palaeogene erosional effects. The Palaeogene strata are largely drift covered, the most extensive outcrops being along the Stour valley southwards from Sudbury and, to the west of it, around Gestingthorpe and also where a low, ill-defined London Clay scarp faces north

near Bulmer. East of the Stour valley, where the drift cover is thicker, the only outcrops are in small windows in the Box valley south-east of Boxford.

The beds lying between the Chalk and the London Clay were termed the Lower London Tertiaries by Prestwich (1852, p.236); the name formerly embraced the Thanet, Woolwich and Reading, and Oldhaven beds. The Oldhaven Beds have been variously grouped with the Woolwich and Reading Beds (e.g. Lake et al., 1986) and the London Clay (e.g. Bristow,1985). In the Sudbury district, as in the Braintree district immediately to the south (Ellison and Lake, 1986), sandy strata which may equate with the Oldhaven Formation are regarded as basal beds of the London Clay. Because of lack of exposure, it has been found impracticable to distinguish the Thanet Beds from the Woolwich and Reading Beds except along the Stour valley; thus the term 'Lower London Tertiaries' is used for the combined formations.

Stratigraphical classification of English Palaeogene rocks, based on foraminiferal, dinoflagellate cyst and nannoplankton zonations, is facilitated by the presence of major hiatuses in the depositional succession between

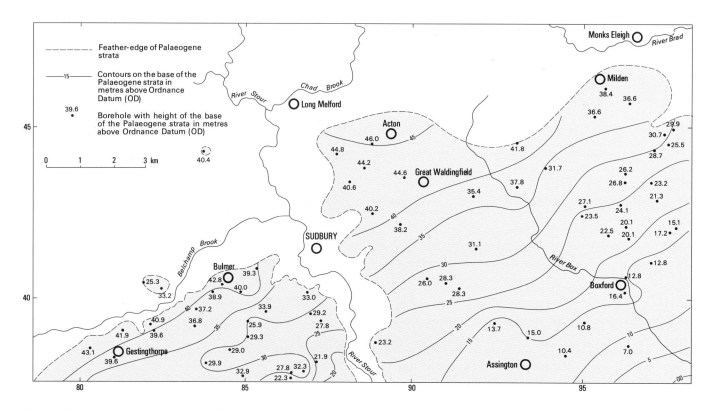

Figure 14   Contours on the base of the Palaeogene strata.

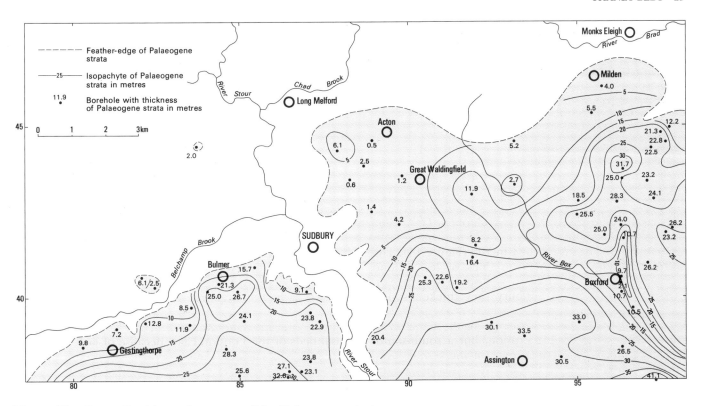

Figure 15 Generalised isopachyte map of the Palaeogene strata.

each of the principal lithostratigraphical divisions. In recent years, work on the nannoplankton, used in conjunction with the magnetostratigraphical reversal sequence and the recognition of widespread correlatable volcanic ash layers (Berggren et al., 1985; Knox, 1990), has led to the classification shown in Table 2.

Table 2 Chronostratigraphical and lithostratigraphical classification of Palaeogene strata.

System/Period	Series	Stages	Formations
	Eocene	Ypresian	London Clay
Palaeogene	Palaeocene	Thanetian	Woolwich and Reading Beds
			Thanet Beds

The Palaeocene–Eocene boundary is drawn at the base of Nannoplankton Zone NP10 and, although stratigraphically diagnostic nannoplankton have not been found in this part of the onshore English succession, the correlation of that boundary with the base of the London Clay is inferred from the recognition of a series of volcanic ash layers, known to be of early NP10 age, in the lowest part of the London Clay in East Anglia, including a section in the Sudbury district (Knox and Ellison, 1979).

The Palaeogene strata of the London Basin, including this district, were deposited in or close to an embayment on the west side of the North Sea Basin. This bay was probably connected westwards with the Atlantic at times. Its northern edge may have lain not far beyond the present-day limit of the Palaeogene outcrop in the region. Ellison (1983) showed that the sandstones constituting the Thanet Beds and the lower part of the Woolwich and Reading Beds are the products of two marine transgressions. Analysis of the heavy minerals in these sandstones by Morton (1982) indicated a northern provenance (from the Scottish Highlands) for those in the Thanet Beds and derivation from the south, possibly the Armorican massif, for those in the glauconitic sands of the Woolwich and Reading Beds. The higher parts of the Woolwich and Reading Beds, comprising finer-grained grey beds with some red and green mottled clays, are thought by Ellison (1983) to have been deposited in a brackish-water lagoon during the regressive phase following the second transgression. The basal transgressive sands of the London Clay represent a third major transgression, but the lithologies and faunas above them indicate a largely low-energy, open-sea, marine environment.

THANET BEDS

Where well-exposed and clearly definable along the sides of the Stour valley around and south of Sudbury, the Thanet Beds are about 5 to 7 m thick and consist of three distinct units: a basal flint pebble to cobble gravel, known as the 'Bullhead Bed'; an overlying green, glauconitic silt or sand; and a top unit of grey and brown to pink silts and sands. Because of poor exposure, a further subdivision of this highest unit into lower pinkish clayey sands and higher finer-grained, grey and buff sands made by Boswell (1929) is no longer evident. He recorded several good pit sections showing all these subdivisions, but only a few of the exposures were still visible at the time of the resurvey (Millward, 1980a; Wilson and Lake, 1983).

Away from the Stour valley, the formation is less easy to define because of the difficulty of recognising the upper boundary in most borehole logs. The magnitude of the variations in thickness of strata assigned to the formation within the district (0.3 to 8.5 m) probably reflects that difficulty. The assigned thicknesses are generally greater west of Sudbury than east of it. Where the Thanet Beds–Woolwich and Reading Beds boundary is confidently defined, it is commonly based on the finer

grain size of the sand fraction in the lower formation. In the broad outcrop of Thanet Beds around Cuckoo Tye, north of Sudbury, the formation is thin (less than 3 m) and consists of green, silty, glauconitic, fine-grained sand, reddened in part, with the basal Bullhead Bed recognised in places.

No macrofossil has been recorded from the Thanet Beds in the district, although marine bivalves and gastropods have been collected elsewhere. However, a few localities sampled during the resurvey have yielded diatoms, radiolaria and foraminifera. The diatoms and radiolaria are mostly indigenous, but most of the foraminifera identified by Mr M J Hughes are reworked Cretaceous species. Foraminifera probably indigenous to the Thanet Beds are agglutinated forms, including *Trochamminoides* sp.

Merriman (in Ellison and Lake, 1986) commented on the widespread presence of volcanigenic material in the

Plate 2   Junction between Thanet Beds and Upper Chalk in an old pit at Ballingdon [8608 4061].

The upper part of the Chalk is very soft. The irregular nature of the junction is attributed to solution collapse in the Chalk. The basal few centimetres of the Thanet Beds consist of brown, clayey, fine-grained sand succeeded by a pebble bed (the Bullhead Bed). The green coating on the flints and the green colour of the overlying fine sands is due to the presence of glauconite. [A13253]

Thanet Beds of the Braintree district to the south. That material includes basic to intermediate lava pyroclasts in the basal beds and zeolites in the overlying sands. In the Sudbury district, heavy minerals indicative of an ash-fall origin (aegerine, apatite and brown amphibole) have been noted in the Thanet Beds of the Great Cornard Borehole [8899 3863] by Mr A C Morton. The aegerine and amphibole were recorded only in the two samples collected from 1.3 m and 2.3 m respectively below the top of the 6.2 m-thick formation. Several of the other heavy minerals found in all or most of the six samples collected through the formation in the borehole (zircon, tourmaline, staurolite, rutile, kyanite, hornblende, garnet, epidote) were also recorded from the Thanet Beds in the district by Boswell (1929, p.18). He remarked that, despite lithological differences between the subdivisions of the formation, the heavy mineral content remained uniform throughout, indicating a continuity of provenance.

The Bullhead Bed is so-named because of the distinctive green-coated, irregularly shaped, nodular, cobble-grade flints it contains. The contents also include rounded, pebble-grade flints and green or brown fine-grained sand. The bed lies on the mostly planar Chalk surface, but locally, as at Ballingdon pit [8608 4061] (Plate 2), that surface is very uneven, probably because of solution collapse of the Chalk. As a result, bedding is disturbed within the overlying Thanet Beds. The Bullhead Bed can be up to 0.45 m thick, but locally it may be represented only by scattered pebbles. The green coatings of flints at Ballingdon are reported by Mr R J Merriman to consist of a randomly interstratified illite–smectite mixed layer mineral which can be classified as glauconite *sensu lato*. A white coating between the green outer skin and the flint itself was found to consist mostly of quartz with traces of calcite.

The green, glauconitic, compact, but generally uncemented clayey silts and sands overlying the Bullhead Bed are commonly less than one metre thick. Glauconitic grains constitute up to 40 per cent of the total rock (Merriman *in* Ellison and Lake, 1986). In boreholes, these deposits are olive-green at the base and greyish green above, but may weather to buff or brown on exposure. The presence of colour-banding, which appears to represent bedding, suggests that at least some of the colour variations are primary. Contemporaneous burrows are commonly filled with material which is greener and more glauconitic than the surrounding strata.

The highest unit of the Thanet Beds is distinguished from the underlying beds principally by colour and locally by grain size. The colours are variously buff, brown, pale grey and olive-grey, but they consistently show a pinkish hue or pink mottling. The mottling may be related to bioturbation. In places, the colour change at the base of the unit coincides with a change of grain size, although this may be either a coarsening or fining. Lithologies include mostly silty sands and sandy silts, generally uncemented and commonly micaceous. Laminated beds may be present, but burrows filled with glauconitic material are also recorded.

DETAILS

One of the cleanest sections in the Thanet Beds seen during the resurvey was at Victoria Pit, Sudbury [8786 4169], where approximately 6 m of pinkish, fine-grained sand and silty sand with some clay passes down into similar but glauconitic material in the basal 66 cm. The Bullhead Bed is well developed, consisting of rounded and knobbly, glauconite-coated flints up to 15 cm across; its base is flat-lying (Millward, 1980a). The most extensive exposures of the formation are probably along the line of old pits between the Newton and Cornard roads in south-east Sudbury [8877 4147 to 8837 4090]. At the northern end, 5.3 m of silty, clayey, fine-grained, uncemented or poorly cemented sand were recorded (Millward, 1980a). The colours range from buff to pale grey (with pale pink mottling) at the top, to olive-grey and grey-green towards the base, with greener, glauconitic material in burrows. The basal 1 cm of the sand is brownish and clayey, with a few glauconite-coated flints, and is underlain by further green-coated flints (i.e. the Bullhead Bed). The latter was also well seen in an old pit [8763 3950] near Middleton, where it consists of a thin band of rolled flints with glauconitic coatings resting on the Chalk and overlain by 0.7 m of glauconitic clayey silt.

WOOLWICH AND READING BEDS

Use of the composite name 'Woolwich and Reading Beds' was recommended by Hester (1965) because strata of the two implied types (the 'Reading type', of largely fluviatile facies, and the 'Woolwich type' of mostly estuarine facies) commonly interdigitate, and thus strata in intermediate areas cannot be classified specifically as one or the other. However, a basal subdivision, comprising marine glauconitic sands and known as the Bottom Bed, is common to both types. In the Sudbury area, mottled clays overlying the Bottom Bed are clearly of the 'Reading type' and were referred to the Reading Beds by Boswell (1929).

The Woolwich and Reading Beds generally overlie the Thanet Beds conformably, but the top of the latter may be burrowed. Red-green mottling may occur throughout and is the single most-used factor in determining the presence of the formation. Without that mottling, the similarity between Bottom Bed lithologies and those of the Thanet Beds contributes to the difficulties of classifying the Palaeogene strata in boreholes. The two subdivisions of the formation, the 'Bottom Bed' and the 'mottled clays', can be distinguished and traced on a regional scale, and their presence is locally recognisable in this district, but their characteristic lithologies interdigitate and blur the distinction between them.

Where the Woolwich and Reading Beds can be distinguished from the Thanet Beds at outcrop along the Stour valley, they thicken southwards from less than 7 m near Sudbury to about 10 m at the southern edge of the district. Elsewhere, beds assigned to the formation in boreholes may be as much as 18 m thick, especially in the south-eastern corner of the district. Although that figure may exaggerate the true thickness, the formation does appear to thicken significantly towards the south. The presence of the extensive Thanet Beds outcrop east of Long Melford and north-east of Sudbury runs counter to Boswell's (1929, p.19) contention that the Woolwich and Reading Beds overlap the older formation to the north, and the 'Reading Beds' outliers, which he recognised as far beyond the main outcrop as Kedington in the upper Stour valley, remain unconfirmed.

'Reading type' Woolwich and Reading Beds are commonly devoid of macrofossils except for a few plants, and no microfossils, other than reworked Cretaceous foraminifera, have been recorded in the district.

The mineralogy of the formation in the adjoining Braintree (223) district was studied by Merriman (in Ellison and Lake, 1986, p.10). The sand-grade content of the Bottom Bed was found to consist mainly of quartz and glauconite, with subsidiary feldspar, muscovite and ilmenite. The predominant green colour is produced by glauconitic sand grains and the pale green clay matrix, in which illite and smectite are the principal constituents. The mottled clay beds are composed largely of illite, kaolinite and smectite, the last particularly concentrated in discrete bands, which are possibly of volcanic origin.

The outcrop of the Woolwich and Reading Beds in southern East Anglia coincides with the maximum concentration of 'sarsens' in the region. They are blocks of hard, silica-cemented sandstones (quartz arenites) found scattered on the present-day land surface. Their distribution and lithology strongly indicate that they are residual clasts which originated within the Bottom Bed of the Woolwich and Reading Beds as discrete bodies of well-cemented sandstone in otherwise uncemented or poorly cemented sand. Boswell (1929, p.22 and fig. 3) mapped their distribution in the Sudbury district and attributed their general absence from areas north-west of the Palaeogene outcrop to an approximate coincidence between the present-day and original extent of the formation.

The greater part of the Bottom Bed consists of sands, clayey in part and commonly laminated and cross-bedded. They are characteristically green and glauconitic, but other colours occur (including cream, red, brown and grey) as colour banding or mottling. The mottled clays are stiff and waxy, with silt laminae in part, and are widely bioturbated. Although red-green mottling is most common, other colours such as purple, lilac, blue-grey, grey-green and orange-brown are recorded. Boswell (1929, p.13) described lenses of coarse sand and gravel, consisting of polished quartz grains, immediately above the base of the unit at Little Cornard [probably about 891 387].

DETAILS

The formation is generally not as well exposed around Sudbury as the Thanet Beds. The best sections were produced by the digging of trial pits, for example that recorded by Millward (1980a) near Ballingdon [8585 4027]:

	Thickness m
HEAD	0.72
WOOLWICH AND READING BEDS	
Clay, smooth, brown with lilac and blue mottling	0.41
Clay, smooth; dark in centre, brecciated	0.17
Sand with flints (probably injected Head)	0.10
Clay, colour-banded; variously orange-brown, ochreous brown, grey and red	0.32
Clay, predominantly bright red, but with colour-banding showing pale grey-green, orange-brown and dark grey; silty bands; flint pebbles at base and 20 cm above	0.58
Silt to fine-grained sand, clayey in part; laminated orange-brown in top 45 cm, grey towards base	1.45

'LOWER LONDON TERTIARIES', UNDIFFERENTIATED

In the outcrops south of the Belchamp Brook valley, and in boreholes on both sides of the lower Stour valley and across to the eastern edge of the district, it has been found impracticable to subdivide the mixed sand, silt and mottled clay sequence between the Chalk and the London Clay into formations. There also may be doubt about distinguishing Tertiary strata from overlying arenaceous drift deposits in some boreholes. Consequently, the overall thicknesses assigned to the 'Lower London Tertiaries' are commonly unreliable. The largest figures are about 21 m for borehole successions on either side of the Box valley (see p. 29), from where the Palaeocene strata appear to thin north-north-westwards towards the basal outcrop, although over 20 m of sand interbedded with grey and 'mottled' clay in a borehole [8704 3950] at Middleton may all belong to the group.

DETAILS

Although the Thanet Beds–Woolwich and Reading Beds boundary could not be mapped in the Bulmer–Gestingthorpe area, some sections can be assigned to one or other formation. Up to 2 m of green glauconitic clay overlying the Chalk in an old pit [8322 4008] near Goldingham Hall may represent basal Thanet Beds. This is probably the pit west of Bulmer church in which Boswell (1929, p.16) described a 'Basement-bed' including flints, overlain by '0 to 3 feet of fine pinkish loam', also referred by him to the Thanet Beds. A trial pit [8487 4058] north-east of Bulmer church exposed, below 1.7 m of head, 0.6 m of clay, grey and sheared in the top half, mottled red and olive-grey below, overlying 1.3 m of similarly mottled, micaceous silt and fine-grained sand with clay and clayey silt laminae; all 1.9 m are interpreted as Woolwich and Reading Beds.

Conjectural outliers of 'Lower London Tertiaries' have been mapped, on borehole evidence, near Foxearth and Belchamp Walter, although neither of them is at the elevation of the inferred basal Tertiary unconformity extrapolated from the main outcrop (see Figure 14). If they are correctly interpreted, these outliers may be relict pockets in depressions (possibly solution hollows) on the Chalk surface. In a borehole [8372 4432] at Foxearth, 2 m of green to grey sand (clayey and silty towards the base, with some flints) overlying the Chalk were referred to the 'Lower London Tertiaries' (Hopson, 1982, p.48). A nearby borehole [8336 4472] proved 10.4 m of beds ('mottled clay', 5.8 m overlying 'green sand', 4.6 m) which might be assigned to the 'group', although a borehole only 30 m away records 'Boulder Clay' resting on Chalk. The Belchamp Walter outlier is inferred from the sequences proved immediately above the Chalk in two boreholes. One of the sequences [8250 4023] was described as 2.5 m of 'clay, very silty; in parts fine sandy, clayey silt; micaceous; race nodules; olive-greyish brown' (in top 0.6 m) overlying further clay, 'mottled green and red, stiff' (Hopson, 1982, p.48).

The greatest thickness of beds assigned to the 'Lower London Tertiaries' in the district is in a borehole [9711 4335] at Castlings Heath, about 2.5 km north of Boxford; it consists of 21.4 m of interbedded green and grey sand and brown silty clay, underlain by Chalk and overlain by silty clay interpreted as London Clay. Thicknesses almost as great are inferred from the logs of boreholes south-west of Boxford. For example, the following sequence was recorded in a borehole [9350 3880] at Assington:

	Thickness m	Depth m
DRIFT	11.60	11.60
LONDON CLAY	12.78	24.38
WOOLWICH AND READING BEDS		
Mottled clay	1.83	26.21
Hard, light sand	1.83	28.04
Green sand and silt	8.54	36.58
Grey silt	4.87	41.45
Brown clay	1.83	43.28
?THANET BEDS		
Hard green sand	1.83	45.11
CHALK		

The inlier of 'Lower London Tertiaries' along the line of the Box valley south-east of Boxford is inferred from the level of their upper boundary in nearby boreholes, one of which [9669 3963] near Stone Street proved, above the Chalk, 10.4 m of green, interbedded sand and sandy clay.

LONDON CLAY

The London Clay outcrop west of the Stour valley closely parallels that of the underlying Palaeocene strata, but east of the Stour, the formation has a much more limited distribution. Overall, it has a gentle regional dip to the south-south-east, comparable to the underlying beds, but in individual exposures an unconformable

Plate 3 London Clay–Woolwich and Reading Beds junction at Cornard brick pit [8899 3853]. A pebbly fill overlies the grey-brown London Clay which rests sharply and unconformably on Woolwich and Reading Beds sand. [A12552]

junction between the Woolwich and Reading Beds and the London Clay can be seen (see Plate 3). The formation is best exposed in the Bulmer–Gestingthorpe area and locally around Sudbury. Except for some small patches along the sides of the Box valley, it is largely obscured by thick drift east of the Stour valley, where the greatest thicknesses (about 16 m) have been recorded, although 20 m or more may be inferred in the southeast of the district.

When compared with the much thicker and more complete sequence in the Braintree district to the south (Ellison and Lake, 1986), it can be inferred that as much as half the thickness of the London Clay in the Sudbury district belongs to what can informally be termed 'basal beds', in which lithologies are laterally and vertically varied, including silts and silty, fine-grained, locally glauconitic sands, as well as silty clays. The base is marked in places by flint and sandstone pebbles, and comminuted shell debris, constituting a local representative of the Suffolk Pebble Bed (King, 1981).

These 'basal beds' probably equate with the 'A' division of the London Clay in the Stock and Hadleigh boreholes of south Essex (Bristow, 1985; Lake et al., 1986) and may include, or totally represent, an Oldhaven Beds correlative.

The strata overlying the 'basal beds' in the district include lithologies more generally characteristic of the London Clay, i.e. stiff, blue-grey, slightly micaceous clay, with scattered 'race', weathering dark grey-brown to yellowish brown; but the distinction between these strata and the 'basal beds' is blurred by the lateral variations within each division. At Bulmer brick pit [833 382], this higher unit is seen to include the volcanic ash bands recognised by Knox (1990) as being indicative of an early Eocene age, and a thin discontinuous cementstone. A correlation of the unit with the lower part of the Harwich Member of the London Clay is suggested by the presence of the ash bands and the cementstone, as well as by general lithology and stratigraphical relationship. The Harwich Member is the basal division of

Plate 4   London Clay with ash bands in Bulmer brick pit [833 382].
A thin veneer of unstratified silty Head overlies bedded silty clays in the lower part of the London Clay.
The pale, yellowish, continuous beds in the top part of the face are weathered ash bands. In the lower part
of the face, the London Clay is rather more silty and is unweathered and olive-coloured at the base. [A12997]

the formation as defined by King (1981) in the north-eastern part of the London Basin.

Biostratigraphical classification of the London Clay is largely based on nannoplankton (coccoliths), planktonic foraminifera and dinoflagellate cysts; only the last have been extracted from the local strata. Assemblages characteristic of the *Apectodinium hyperacanthum* Zone, which spans the basal London Clay boundary and includes all of the Harwich Member, were identified by Dr R Harland in samples collected from the Bulmer Tye (Butler's Hall Farm) Borehole [8340 3791]. The only other identifiable fossils collected in the district are bivalves from the cementstone bed at Bulmer brickpit (see below).

The clay mineralogy of the London Clay in the adjacent Braintree district, when compared with that of the underlying Woolwich and Reading Beds, shows an increase in the proportion of smectite and a decrease of kaolinite, the <2 μm fraction of the London Clay typically comprising illite 44 to 61 per cent, smectite 26 to 41 per cent, and kaolinite 13 to 15 per cent (Merriman, in Ellison and Lake, 1986, p.13). Sand-grade grains in the

'basal beds' include quartz, feldspar, muscovite, chlorite and glauconite. The ash bands of the Harwich Member are clearly visible in the field only where they occur in non-bioturbated, well-laminated, weathered sections. In southern East Anglia the ash is mostly converted to smectite with a variable pyrite content; in effect to bentonites, except where, as Knox and Ellison (1979) point out, 'such alteration has been inhibited by early calcite cementation'. However, thin sections of the cementstone at Bulmer were examined by Mr R K Harrison and found to consist of a partly pseudobrecciated micritic calcite matrix with detrital sand-grade clasts of quartz, chert, glauconite, chlorite, biotite, muscovite, phosphatic spicules and zircon. No particles of undoubted volcanic origin were detected.

DETAILS

The main section at Bulmer, exposed at the time of the resurvey, is illustrated in Plates 4 and 5. The cementstone yielded the bivalves *Arctica* sp., *Astarte* sp. and a doubtful *Cytodaria* sp. A

Plate 5 Fossiliferous cementstone bed in London Clay at Bulmer brick pit [8330 3823]. Pale mottled silts and clayey silts with cementstone at base overlying bioturbated clayey silt. [A12522]

further 2.5 m of London Clay, stratigraphically below that exposed in the main face and consisting of grey-brown clayey silt, were exposed in a pit 25 m to the west. Strata lying within a few metres of the base of the formation were exposed in a trial pit [8940 4067] west of Abbas Hall, Great Cornard:

	Thickness m
Soil	0.2
Sand, fine-grained, silty, clayey, micaceous, with scattered flints in the top 0.7 m; pale brown, becoming mottled pale grey towards gradational base	1.1
Clay, very silty, brown and pale grey mottled, with scattered mica	0.7

No clean sections of London Clay were seen at surface farther east. One of the thickest London Clay sequences described there is in a borehole [9554 3867] south of Turk's Hall, Boxford (Hopson, 1981):

	Thickness m	*Depth* m
Soil and sandy gravel	2.4	2.4
LONDON CLAY		
Clay, silty, sandy, micaceous, orange-brown	4.2	6.6
Clay, silty, dark grey	0.4	7.0

SIX

Pliocene–Pleistocene solid: Crag

The term Crag is here used for a lithostratigraphical group covering all the Pliocene–Pleistocene marine sediments in the district. A recent paper by Mathers and Zalasiewicz (1988) provides a regional analysis of several aspects of Crag geology and references to previous studies. They advocate formational status to divisions of the 'Crags', including a lower 'Red Crag' and upper 'Norwich Crag', but it has not been feasible to map the formations separately in this district.

The combined Red and Norwich Crag is commonly assigned to the Lower Pleistocene on the basis of local climatic change (indicated by the entry of cold-water molluscan faunas) and mapping convenience (Boswell, 1952; Mitchell et al., 1973). Nevertheless, it has long been acknowledged that the position of the internationally recognised Pliocene–Pleistocene boundary in Italy, which is now formally defined on biostratigraphical and palaeomagnetic evidence (Aguirre and Pasini, 1985), implies that much of the Red and Norwich Crag sequence is of Pliocene age.

The Red Crag is thought to be of Pre-Ludhamian, Ludhamian and Thurnian age (Gibbard and Zalasiewicz, 1988). It is mainly composed of poorly sorted, cross-bedded, coarse-grained ferruginous sands and is shelly and/or gravelly in part, with a flint conglomerate commonly present at the base. The flints are probably derived from the Thanet Beds (Merriman, 1983). The shells are mostly shallow-water bivalves and gastropods. The overlying Norwich Crag, probably of Bramertonian age, is also a shallow-water deposit but relates to a less energetic environment, because it consists mainly of fine-grained sand with associated silt and clay; it includes the Chillesford Sand Member, which may indeed represent it in this district. Both formations are at least partly glauconitic and this is reflected in many old borehole references to 'green sand', although the Red Crag, in particular, is commonly thoroughly oxidised at surface. Phosphatic nodules may also be present.

A generalised regional distribution of Crag in Suffolk and Essex is illustrated by Mathers and Zalasiewicz (1988, fig. 2), but they postulate a more extensive distribution in the Sudbury district than has been inferred during the recent survey. They demonstrate that recognition of Crag and of the distinction between Red Crag and Norwich Crag is difficult without detailed sedimentological analysis. There is no natural exposure and only a few quarry sections in the district, and such an analysis has only been possible where plentiful borehole material has been available. However, the available evidence proves that the present distribution of Crag is mainly an effect of preglacial erosion acting on a formerly continuous cover of sediment. Mathers and Zalasiewicz (1988) draw attention to the fact that the present landward Crag sequence appears to be incomplete with reference to its counterpart in the southern North Sea. This implies phases of relative isostatic uplift during Crag sedimentation, prior to a more pronounced post-depositional uplift giving a general easterly gradient to Crag deposits as a whole (Mathers and Zalasiewicz, 1988, figs. 1 and 11).

The main Crag outcrop is in the north-eastern quadrant of the district, where there are only subordinate 'windows' of Chalk through an otherwise continuous Crag cover. Elsewhere, the Crag is discontinuous, and increasingly so towards the west and north-west: only isolated relics remain west of Cavendish and it appears to be absent west of Stansfield and north of Kedington, although borehole evidence is particularly sparse in that area.

Borehole data prove that the thickest remaining deposits of Crag in the district lie in a north-east-trending trough on the Chalk surface, known either as the Kettlebaston Basin (Bristow, 1983) or the Stradbroke–Sudbury basin (Mathers and Zalasiewicz, 1988). Bristow's figures 2 and 3 show contours on the Crag base down to below 10 m above OD and Crag thicknesses over 30 m respectively in the basin within the eastern half of the district. It is clear that the thickest Crag and the deepest part of the basin lie close to its eastern margin, which is interpreted by Bristow as a normal fault defining the limit of a graben (see Figure 16). Such an explanation certainly might account for the undulose base of the basin, which is otherwise hard to explain in a nonglacial environment. Another, although less likely possibility, would be an origin involving extensive cavern collapse during karstic subaerial erosion of Chalk prior to submergence beneath the Crag sea.

DETAILS

The thickest recorded occurrence of Crag sediment is in a borehole close to the eastern margin of the district at Hitcham House [9780 5069], where 41.76 m of 'sand and shell' were noted below 6.10 m of 'hard yellow sand'; the latter may also be Crag, but is more likely to be Kesgrave Sands and Gravels. The BGS Hitcham Borehole in the same area [9829 5113] (Bristow, 1983, p.8) proved 13 m of Kesgrave Sands and Gravels on 21.7 m of Crag, of which the top 17 m are of Norwich Crag type and the remainder, including a basal gravel resting on Chalk, is of Red Crag affinity. Similar sequences are recorded in boreholes south of Lavenham, towards the south-western end of the Kettlebaston Basin. Here, the thickest occurrence is at Elm Tree Farm, near Lavenham [9162 4723], where 4.57 m of probable Kesgrave Sands and Gravels overlie 36.57 m of Crag. In a neighbouring borehole [9187 4735], where 30.48 m of Crag are inferred to be present, the topmost 10.06 m of sand are variously described as loamy or having associated clay, whereas the lower beds are referred to as 'sand' with 'flint gravel' resting on the underlying Chalk.

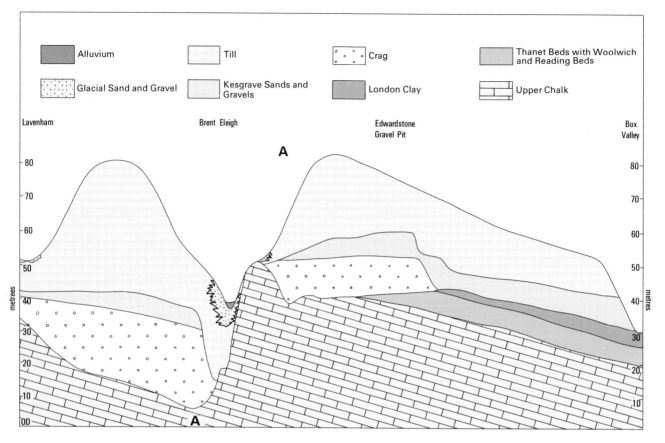

Figure 16   Diagrammatic section showing inferred stratigraphical relationships along an approximately north to south line in the east-central part of the district (TL 94 NW and SW). The position of the postulated 'Kettlebaston Fault' of Bristow (1983) is indicated by the letters A–A'.

Crag exposures in the district are now mostly limited to parts of the overburden in old chalk pits at Sudbury and Monks Eleigh, which both show the basal few metres of probable Red Crag. Details of the previously more abundant sections visible around Sudbury were provided by Boswell (1929). Millward (1980a) recorded 2.45 m of Red Crag overlying Thanet Beds in an old pit [8869 4139] east of Sudbury town centre. It comprises fine- to coarse-grained, partly cross-bedded sand with some com-minuted bivalve shells and sparse quartz pebbles. The sharp regular base is overlain by well-polished flint pebbles. Millward also referred to a small outcrop of probable Crag between Long Melford and Stanstead, in which an exposure at Cranmore Green [8504 4778] revealed 0.8 m of micaceous, coarse-grained, yellow-brown to rusty brown sand. The deposit consists of well-polished rounded quartz grains with scattered quartzite pebbles, interbedded with subsidiary, brown mottled silty and sandy clay.

SEVEN

Structure

PRE-MESOZOIC STRUCTURES

The oldest structures in the district are related to the north-west to south-east-trending, broad, low-amplitude variations shown in the aeromagnetic anomaly map (Figure 4). Their age is unknown but they are much deeper than, and thus are likely to predate structures of similar, (even identical) alignment indicated by the Bouguer gravity anomaly map and thought to be of late Caledonian age. The aeromagnetic data also suggest structures with a north-north-east trend, including an anticline near Lavenham (Figure 17). Its alignment is paralleled by an inferred synclinal structure within deep 'basement' rocks in the north-central part of the district.

Faults inferred from both the aeromagnetic and gravity evidence include a north-north-west-trending one near the south-west corner of the district (Figure 17), but the evidence is contradictory on the direction of throw, as noted also by Linsser (1968). The gravity evidence suggests that low-density Devonian sediments are preserved on the downthrown, (south-west) side, while the aeromagnetic evidence suggests that magnetic basement at a depth of 4 km+ is preserved at a lower level on the same side of this fault; but in structural terms, on the upthrow side. Nevertheless, although the coincidence of strong gradients in both data sets cannot be related to a single coeval source, the faults inferred from them may be related, perhaps in different tectonic episodes; namely, an older phase associated with deeper and older, magnetic igneous basement within the pre-Silurian rocks and a later reactivation phase in late Caledonian times when the 'tectonic' basins of Devonian sediments were being

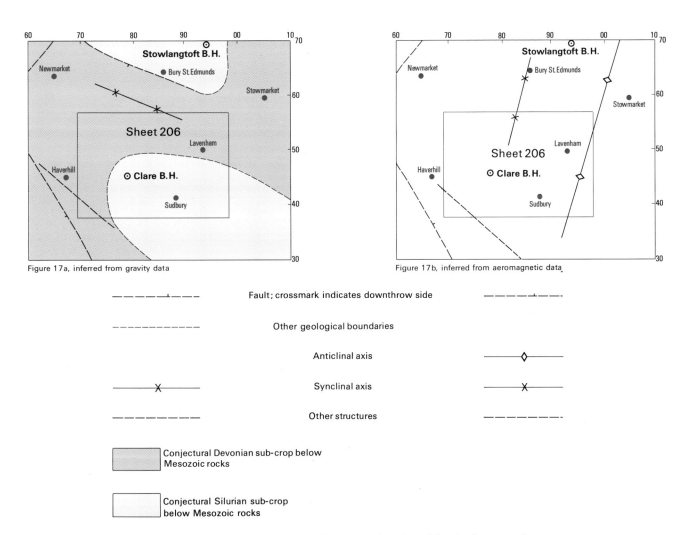

Figure 17a, inferred from gravity data

Figure 17b, inferred from aeromagnetic data

— — — — — ⊥ — — — Fault; crossmark indicates downthrow side   — — — — — ⊥ — —·

— — — — — — — — — Other geological boundaries

—————◇————— Anticlinal axis   —————◇—————

—————✕————— Synclinal axis   —————✕—————

— — — — — — — — Other structures   — — — — — — — —

Conjectural Devonian sub-crop below Mesozoic rocks

Conjectural Silurian sub-crop below Mesozoic rocks

Figure 17 Conjectural geological structures in the 'basement' rocks of the Sudbury region.

formed. This putative fault extends south-eastwards via the Colchester district to the east coast and separates Silurian rocks to the north-east, proved in the Harwich, Weeley and Stutton boreholes, from a thick sequence of low-density Devonian rocks to the south-west, proved in several water boreholes in the London area and at Canvey Island.

A second gravity gradient near the south-west corner of the district, which appears to merge with the other towards the west, may indicate the local boundary between Silurian and Devonian rocks. However, this minor gradient of about 0.7 mGal km^{-1} (the other is 2.1 mGal km^{-1}) is probably not the expression of a faulted junction.

The gravity anomaly map also suggests the presence of WNW-ESE-aligned structures within the basement rocks near the northern edge of the district. The evidence comprises a gravity low bounded on its northern side by a steep gradient within the Bury St Edmunds district (Bristow, 1990). The structures are taken to represent a synclinal trough with thick Devonian sediments, associated with a major fault of similar trend; the latter would separate the inferred Devonian strata from Silurian rocks to the north which were proved in the Stowlangtoft Borehole. The southern edge of the gravity low has a more east to west alignment and coincides approximately with a weak structure in the overlying Chalk (see Figure 10). However, if the gravity effect of the Mesozoic rocks is removed from the observed Bouguer anomaly field, east–west-trending residual closures of 0.85 to 2.4 mGal remain; these values are too great to be accounted for by the 'tunnel valley' in the same area. The residual gravity gradient is therefore taken to mark the southern edge of the synclinal trough to the north of Clare.

Basement structures with a north-east–south-west or NNE–SSW trend are also suggested by the Bouguer gravity data. This alignment coincides with structures observed in the Cretaceous rocks of both the Cambridge district to the north-west (Worssam and Taylor, 1969) and the western part of the Sudbury district. However, the basement structures with a north-north-east trend, near Lavenham and in the north-central part of the district suggested by the aeromagnetic data (see above), do not appear to affect the overlying Mesozoic rocks.

MESOZOIC AND POST-MESOZOIC STRUCTURES

The Cretaceous strata show a general east to south-east regional dip of about 1 in 200, but slight variations occur in the Chalk outcrop, with structural trends ranging between east–west and NNE–SSW (Figure 10). Some of these minor structures may be aligned with those affecting the basement rocks (see above). For instance, the north-north-east-trending basement structure in the north-west corner of the region shown in Figure 17 parallels a strike-parallel steepening of the Chalk dip in the north-west corner of the Sudbury district. A diversion of the Chalk structure contours along the line of the middle Stour valley (Figure 10) appears to indicate an east to west—aligned flexure or fault. This postulated structure closely parallels the inferred southern edge of the synclinal trough filled with Devonian sediments in the northern half of the district.

The Tertiary strata have a gentle regional south-south-eastwards dip and form part of the northern limb of the London Basin syncline (Figure 14). Wooldridge (1923) described minor structures affecting the Tertiary rocks elsewhere on the northern side of the London Basin but none has been recognised in this district.

Bristow (1983) identified several north-east–south-west structures across central Suffolk on the basis of variations in the thickness of the Crag recorded in boreholes. They included the Kettlebaston Fault, which he showed extending into the eastern part of the Sudbury district and crossing the area between the Stour and Brad valleys near Little Waldingfield. The spatial relationships of the Tertiary and Quaternary strata there are diagrammatically illustrated in Figure 16, which appears to indicate the presence of a fault because of the relative height of Crag deposits to the north and south of the buried valley in the Chalk surface. There is, however, no data concerning the relative ages of the two sets of Crag deposits; the southern set may be younger than the northern, rather than fault-displaced rocks of the same age. Mathers and Zalasiewicz (1988) drew attention to the lack of direct evidence for major Quaternary faulting in this region, and to other possible causes of strong relief on Chalk surfaces.

EIGHT

Pre-Anglian drift deposits: Kesgrave Sands and Gravels

The Kesgrave Sands and Gravels are distinctive in composition, differing from both the underlying Crag and the later Glacial Sands and Gravels. The Glacial Sands and Gravels directly overlie the Kesgrave Sands and Gravels locally, but are commonly separated from them by till.

Typically, the Kesgrave Sands and Gravels consist of 'clean', medium- to coarse-grained quartzose sands unpredictably interbedded with gravels rich in well-rounded, commonly rather small pebbles, of which a high proportion are quartzose; white vein-quartz and black chert are particularly prominent. These deposits are distinguished from the underlying upper Crag beds mainly by their coarser grain size, and from the overlying glacial deposits by the high degree of rounding of their pebble fraction, with its absence of local material, especially Chalk. Although the deposit is typically 'clean' throughout (but see comments on p.56), the top metre is locally iron-stained and clayey, with physical disruption of the bedding. This has been described as a 'sol lessivé' and attributed to two stages of postdepositional, preglacial pedogenesis (Rose and Allen, 1977; Rose et al., 1985).

The Kesgrave Sands and Gravels are regarded as fluvial deposits laid down during the Beestonian cold stage of the Middle Pleistocene. The parent river, or rivers, are assumed to have flowed north-eastwards across this region (Rose et al, 1976), which lay to the south of a contemporary ice sheet. The postulated fluvial origin of these deposits accords with their occurrence as terrace-like tracts at various elevations (see Figure 16). The complete suite of terraces may have suffered some of the isostatic uplift that affected the earlier Crag, although they were deposited as much as a million years later than the youngest Crag. Farther south, Hey (1980) divided the formation into high-level (Westland Green Member) and low-level (unnamed) gravels; these subdivisions have not been recognised in the Sudbury district, where the deposits appear to lie within Hey's mapped outcrop of the Westland Green Member.

It seems probable that Kesgrave Sands and Gravels covered all but the north-western quarter of this district, but they have been removed by glacial erosion from buried channels (Figure 19). Exposures are restricted, except for a small inlier 1 km north-west of Glemsford, to valley sides in the south-east of the district where the cover of glacial deposits is thinnest. The widespread outcrops shown on the map to the south and south-west of Sudbury probably exaggerate the extent of the Kesgrave Sands and Gravels because they include representatives of the Barham Sands and Gravels (Rose and Allen, 1977). These later deposits are commonly reddened and are not always readily differentiated in the field from the 'sol lessivé' at the top of the Kesgrave Sands and Gravels.

The maximum recorded thickness of the formation in the district is an unbottomed 17.2 m in a borehole at Rob's Farm, Cavendish [7962 4839]. The deposits in this particular area (Grid square TL74, including sections around Clare, Birdbrook and Poslingford) are uncharacteristically clayey, though they still feature the typical rounded quartzose pebbles. These are the westernmost occurrences of the formation identified in the district. Hopson (1982) considers that a clay-rich variant of Kesgrave Sands and Gravels, with a strongly unimodal particle-size distribution, may predate the more typical clean gravelly type in the area around Sudbury. South and south-west of Sudbury, where the formation is at or near the surface over wide areas, the maximum thickness recorded is 15.9 m, with a quoted average (Marks and Merritt, 1981) of 7.3 m.

North of Sudbury, a thickness of 12.2 m has been recorded beyond the Stour buried valley near Bridge Street, and east of Sudbury up to 9.2 m are present in the well-exposed valley sides around Boxford. About 6 m of Kesgrave Sands and Gravels are inferred to underlie much of the Lavenham area. To the north and west of Lavenham, the distribution and thickness are conjectural, but much of the ground in the Hartest, Cockfield and Brettenham area is probably underlain by Kesgrave Sands and Gravels.

DETAILS

The only fresh sections of Kesgrave Sands and Gravels seen in this district were at Edwardstone gravel pit (Lynn's Hall Quarry) [935 432], which was one of the localities cited by Rose et al. (1976) in their description of the formation and the palaeosol marking its top. About 12 m of till overburden overlie 7 to 8 m of Kesgrave Sands and Gravels, comprising mixed sand- to gravel-dominant assemblages, which are generally flat-bedded. Some individual beds show cross-bedding, however, and there are also channel cross-section structures, best seen on north–south faces, which indicate preferred west to east channel orientation. The top metre showed physical disruption and some apparent mixing with overlying till, possibly a local expression of the Barham 'Sol Lessivé' (Rose et al., 1985), although no obvious reddening was observed. The base of the Kesgrave Sands and Gravels was not seen, but it reportedly lies on 2 to 4 m of fine-grained white sand on fine-grained rusty brown sand, both conjecturally regarded as part of the Norwich Crag.

NINE

Anglian drift deposits

Middle Pleistocene glaciation was the main influence in shaping the present-day landscape of southern East Anglia and was the source of most of its surface deposits. This much has been agreed since the term 'Middle Pleistocene' was first used, but how many ice advances were involved and to which stage or stages they should be assigned have remained matters of dispute. Previous proposals by Baden-Powell (1948), West and Donner (1956) and Clayton (1957) that the local tills were deposited by separate advances during both the Anglian and Wolstonian have been superseded by the thesis that only one glacial episode was involved. Bristow and Cox (1973) referred this to the Wolstonian, but Perrin et al. (1979) suggested that the Anglian was more likely and the latter has now become the established view.

The Anglian ice sheet appears to have advanced from north-west to south-east across the region and, at its maximum extent, to have formed a front parallel to and slightly inland from the north Essex–south Suffolk coast. Its southward passage over eastern England across the broad outcrops of Upper Jurassic clays and Upper Cretaceous chalk is reflected in the composition of the associ-

ated glacigenic deposits. These consist of proglacial and englacial sands, gravels and silts, as well as large thicknesses of till (Figure 18).

The proglacial deposits are thought to include the Barham Sands and Gravels, which underlie and extend beyond the till. Their composition suggests a mixed provenance from both the Anglian ice and the Kesgrave Sands and Gravels. Although the Barham Sands and Gravels have been recognised within the district (Rose et al., 1976; Hopson, 1982), overall it has been found impossible to distinguish them from the subglacially deposited sands and gravels, which also, in part, underlie the till.

Water-sorted deposits, consisting of silts, sands and gravels, occur below, within and above the till, and have a composition, including much chalk and flint, of sufficiently close affinity to it, to indicate a similar provenance. They are particularly concentrated along the lines of modern valleys, demonstrating the continuity of the drainage pattern since glacial and probably preglacial time. The subglacial, englacial and supraglacial drainage appears to have been generally south-easterly towards the

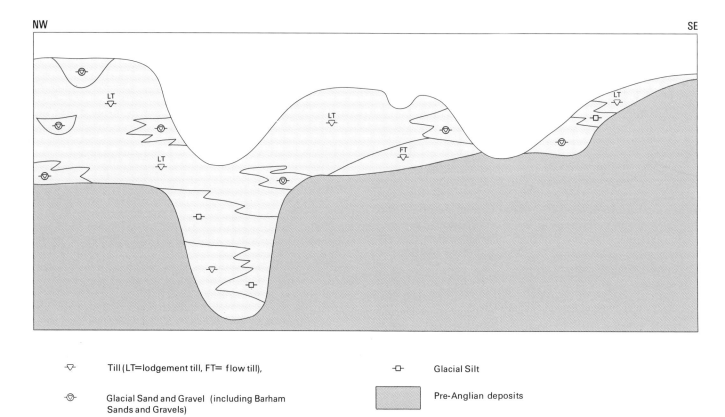

Figure 18 Diagrammatic section showing spatial relationships between the different Anglian glacigenic deposits in the south-central parts of the district.

ice-front, but the distribution of sand and gravel, and of subdrift channels, also suggests many transverse streams. Although some of the sand and gravel may have been deposited below the active ice sheet, most of the water-laid deposits, especially the silts, probably originated during the waning of the ice sheet.

The tills, which in the past have been variously sub-divided and named, are now all probably referable to the Lowestoft Till. They include considerable variations in lithology and fabric although, for the greater part, the old term 'Chalky Boulder Clay' is appropriate. Most of the till appears to be a structureless lodgement till, the Blakenham Till of Rose and Allen (1977), but in some eastern and south-eastern parts of the district, the lodgement till overlies horizontally banded chalky and flinty clays, which are taken to be flow till, proba-bly equivalent to the Creeting Till of Allen (1984). In some north-eastern parts of the district, an Anglian till (the Starston Till of Lawson,1982) predating the Low-estoft Till, may be present, but the evidence is negli-gible.

Although glacial deposition was the most visible effect of the Anglian ice, glacial erosion also had a major impact on the district. Several drift-filled valleys, including one cut down to at least 112 m below OD along the line of the middle Stour, could only have been eroded subglacially, either by the ice itself or, more probably, by a subglacial stream under considerable hydrostatic pressure.

DRIFT-FILLED CHANNELS

Channels cut into bedrock and filled with Drift deposits, including glacial sand and gravel, till and laminated clays and silts, are common in an arc along the largely till-covered Chalk subcrop of East Anglia (Woodland, 1970). The channels mostly follow the line of modern valleys, although they differ from a subaerial drainage system in two important ways. Firstly, their longitudinal profiles appear not to slope uniformly in one direction and, in places, show marked overdeepening, even to lev-els considerably below present sea-level. Secondly, al-though some of the channels form a dendritic pattern in plan, others are arranged in a reticulate grid, as, for ex-ample, west of Norwich. Both factors indicate that pro-cesses other than normal subaerial fluviatile erosion were involved.

Buried channels like those of East Anglia occur else-where in areas of glacial deposition across northern Europe and have been called 'tunnel-valleys' in Den-mark and 'Rinnentäler' in north Germany. Three gener-ations of tunnel-valleys have been recognised below the North Sea and are assigned to the Anglian, Wolstonian and Devensian stages (Cameron et al., 1987). Ehlers et al. (1984) attributed these channels to erosion by sub-glacial streams and their overdeepening to the great hy-drostatic pressure possible in such environments. Cox (1985), however, considered that the thalwegs of some East Anglian tunnel valleys were the products of low in-terglacial sea levels.

Several tunnel valleys have been detected in this dis-trict (see Figure 19) of which the longest and deepest follows the line of the middle Stour valley between Wixoe and Long Melford. A depth to rockhead of 154 m proved by the Clare Borehole [7834 4536] indicates a channel bottom at least 112 m below sea level, making it the deepest known of the East Anglian buried channels. Its general course was delineated by a gravity survey, and confirmed locally by resistivity transects (Barker and Harker, 1984) and by many boreholes (Figure 20). The geophysical surveys indicate remarkably steep channel sides; at least 45° in places. Both in their 1984 paper and subsequently in response to Cornwell (1986), Barker and Harker acknowledged considerable imprecision in their channel depth predictions because of the varied nature of the channel fill and the similarity between the density of its sand and gravel component and that of the chalk bed-rock. The longitudinal profile shown on Fig-ure 20 is thus questionable, as must be the existence of two parallel channels within the tunnel valley between Clare and Liston, separated by a chalk ridge up to 70 m high, a model suggested by Barker and Harker. Detailed gravity surveys, in combination with EM31 resistivity sur-veys, were also used to determine the extent of tunnel-valley deposits in the adjacent Bury St Edmunds district (Cornwell, 1985).

In common with other tunnel valleys in East Anglia described by Woodland (1970), the proportion of sand and gravel in the fill of the Stour buried channel ap-pears to increase in a downstream (i.e. in an easterly direction). The change from a largely till or laminated clay fill, to one with a significant percentage of sand and gravel, approximates to the position of a rise in the lon-gitudinal profile indicated by the gravity survey. Bore-holes also show considerable lateral variations in the fill content, with valley-side wedges of lacustrine silts and clays, and interdigitations of till with sand and gravel, as well as large chalk rafts. A chalk raft 21 m thick was proved by the Anglian Water Authority borehole at Claredown Farm [7778 4472]. Along the centre line of the channel, the glacigenic fill is entirely covered by younger fluviatile and soliflucted deposits.

The course of the Stour tunnel valley extends east-wards beyond the southward turn of the modern Stour valley at Long Melford but is in line with that of a buried channel below the River Brad at Monks Eleigh. This suggests a west to east subglacial drainage link between them although evidence for a tunnel valley in that area is inconclusive. Other shallow buried valleys have been proved under the upper Glem (possibly extending east-wards beyond Hartest), the upper Stour near Keding-ton, the Colne at Great Yeldham and the valleys of the River Box and Belchamp Brook. The tunnel valleys below the River Lark catchment area described by Bristow (1990) probably just enter this district near Stanningfield.

Because these buried channels cut through Anglian till, and no post-Anglian glaciation is now considered to have reached this district, they and their fill of glacigenic sediments must be regarded as of Anglian

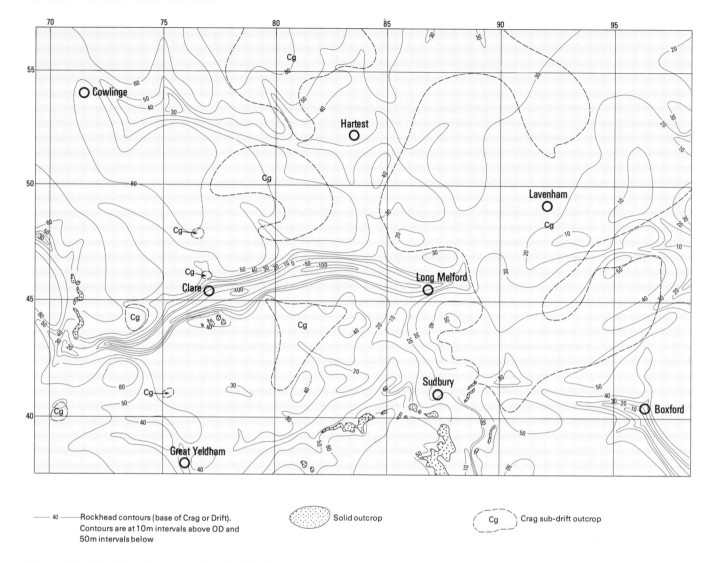

Rockhead contours (base of Crag or Drift). Contours are at 10m intervals above OD and 50m intervals below

Solid outcrop

Cg   Crag sub-drift outcrop

Figure 19   Rockhead contours in the district.

age. However, it is possible that periglacial deposits of later cold stages are represented in the upper part of the channel fill.

DETAILS

Generalised graphic logs of boreholes along the line of the Stour tunnel-valley are shown in Figure 20. A more detailed description of the drift deposits proved in the Clare Borehole (No. 9 in Figure 20) is as follows:

	Thickness m	Depth m
ALLUVIUM AND RIVER TERRACE DEPOSITS		
Sand, orange-brown, clayey	1.0	1.0
Sand and gravel, orange-brown, clayey	1.7	2.7
TILL		
Clay, pale grey, silty, sandy in part, with flint, chalk, quartzite and vein quartz pebbles; very		
chalky in basal metre	33.3	36.0
GLACIAL SAND AND GRAVEL		
Sand, medium- to coarse-grained, chalky and gravelly; gravel mostly flint, white patinated, round and angular	4.2	40.2
TILL		
Clay, mostly grey but grey-brown at top; variably chalky; rare flints	18.8	59.0
GLACIAL SAND AND GRAVEL		
Gravel; mostly flint pebbles	2.5	61.5
GLACIAL SILT		
Silt and clay, soft; scattered flint pebbles	5.5	67.0
TILL		
Clay, grey-brown to grey, silty; finely sandy in part, small chalk pellets common, especially towards base; some angular flints	9.0	76.0
GLACIAL SILT		
Silt and clay, soft; some chalk throughout	29.0	105.0

	Thickness m	Depth m

Till
Clay, dark grey, chalky   1.5   106.5

Glacial Silt
Clay and silt, medium grey, soft   5.5   112.0

Till
Clay, grey, grey-brown in part, sandy in part;
 scattered chalk and flint   18.0   130.0

Glacial Silt
Clay and silt, soft   2.0   132.0

Till
Clay, grey-brown; scattered chalk and flint   7.0   139.0

Glacial Silt
Clay and silt, soft   11.0   150.0

Till
Clay, grey-brown, silty, sandy; chalk fragments
 common   4.0   154.0

Middle Chalk

As in the main Stour buried channel, the drift deposits filling the other tunnel valleys in the district consist largely of glacigenic sediments, especially till, commonly covered in part by younger Head and/or Alluvium. A borehole [7592 3785] at Great Yeldham showed:

	Thickness m	Depth m
Glacial Sand and Gravel	11.6	11.6
Till	9.8	21.4
Glacial Sand and Gravel	1.4	22.8
Till	9.8	32.5
Gravel (?glacial or *Kesgrave Sands and Gravels*)	1.7	34.1

Upper Chalk

Rockhead was at 20.8 m above OD.

The buried channel below the line of the River Box was proved by a borehole at Boxford [9622 4043] in which 22.85 m of Head and Till were recorded above rockhead at 6.1 m above OD. Another [9715 3875] at Peyton Hall proved:

	Thickness m	Depth m
Head	3.7	3.7
Glacial Silt	4.0	7.7
Till Silt, greyish green and grey, clayey, with some very clayey seams containing coarse sand-grade chalk clasts	11.3	19.0

The base of this sequence was at 4 m above OD.

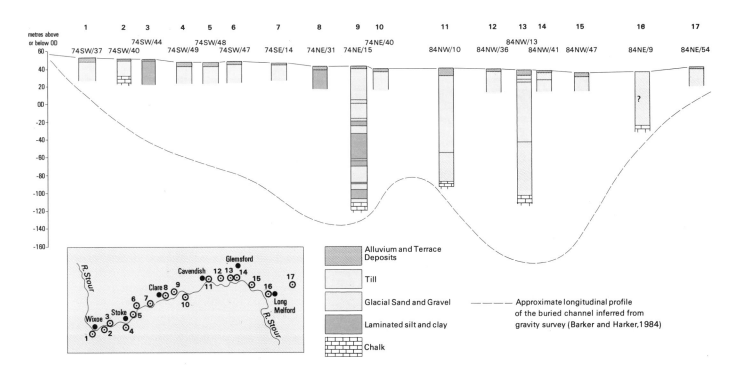

Figure 20   Borehole sections along the Stour tunnel valley.
1. The boreholes are not necessarily on the central line of the buried valley.
2. The figures 1–17 above each log refer to positions on the inset map. The numbers in brackets below them are those of the BGS records system in which each is preceded by TL.

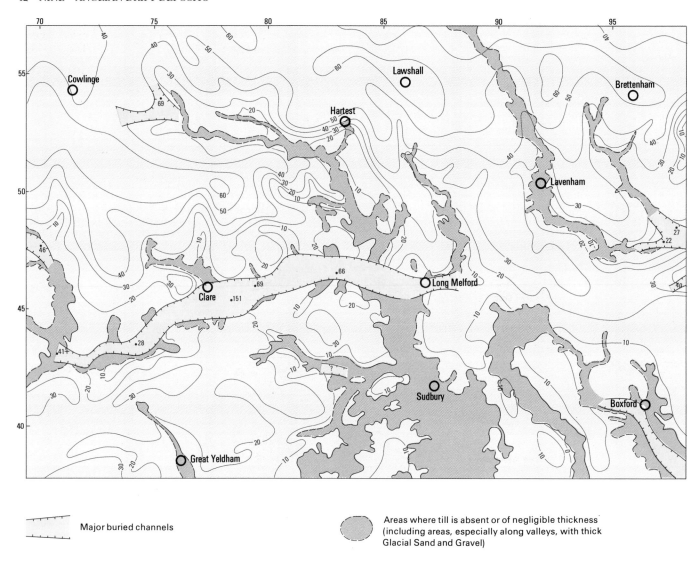

Major buried channels

Areas where till is absent or of negligible thickness (including areas, especially along valleys, with thick Glacial Sand and Gravel)

Figure 21 Generalised thickness of till. Isopachyte lines at 10 m intervals except in the deep sub-Drift buried channels, where till thicknesses in selected boreholes are given. The figures indicated throughout are for total thicknesses of beds between the highest and lowest till recorded and may, therefore, include glacigenic gravel, sand and silt.

TILL

The Lowestoft Till covers most of the district, both in the plateau-like interfluve areas where, together with subordinate sand and gravel, it forms a subhorizontal sheet, and in the tunnel valleys which it partly fills. The plateau-forming sheet is thickest, up to more than 60 m, in the northern half of the district (Figure 21). In the southern half, its thickness exceeds 30 m west of the lower Stour, and reaches about 15 m east of the river. Till is absent from parts of the deeper modern valleys, such as those of the River Box and lower Stour, which are incised through the plateau sheet, and from limited areas where the entire Anglian glacigenic succession consists of sand and gravel, such as the strip of land running from Great Waldingfield through Newton to Assington.

In general, the thickness of the till sheet varies with the present-day height of the land, which is largely a product of post-Anglian planation and subsequent fluviatile incision. In contrast, the level of its base is much more uniform (see Figure 22). There are two principal departures from this pattern: the till thicknesses up to 60 m and more around Lawshall and Brettenham, despite the relatively low plateau there, are due to a depression in the subtill surface, and till is absent or thin on a sub-Drift ridge of Tertiary bedrock south-west of Sudbury.

The unweathered till is a chalky, bluish grey to brown, variably silty, sandy and stony clay. Perrin et al. (1979) conducted mineralogical and particle-size analyses of 382 Lowestoft Till matrix samples collected over a large part of eastern England, including the Sudbury district. They demonstrated that the Lowestoft Till and related drift de-

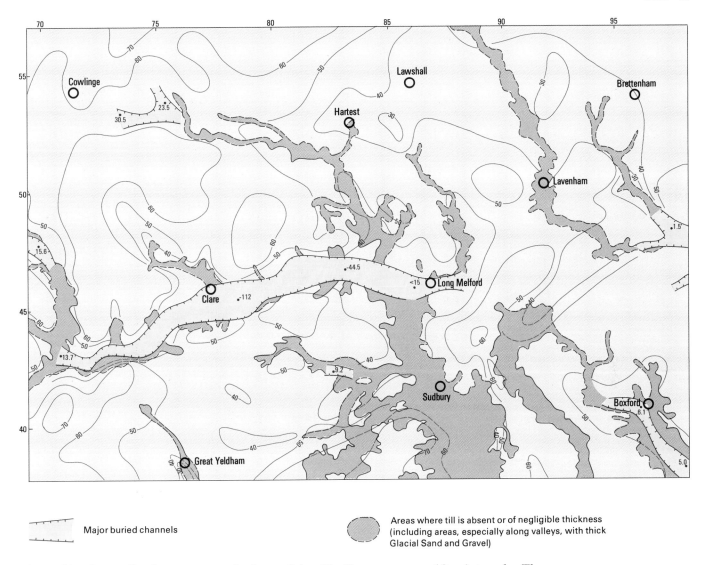

Major buried channels

Areas where till is absent or of negligible thickness (including areas, especially along valleys, with thick Glacial Sand and Gravel)

Figure 22   Generalised contours on the base of the till. Contours are at 10 m intervals. They are not shown in the deep sub-Drift buried channels where till distribution is very irregular and selected base of till levels are shown instead.

posits differ from the Cromer Till and associated deposits in northern and eastern Norfolk in their heavy mineral assemblages and in the low sand-grade content of the insoluble residue. The Lowestoft Till matrix was shown to have an average calcium carbonate content of 40 per cent.

The clay fraction mineral assemblage of the 'Chalky Boulder Clay' consists of 'mica and kaolinite, with variable smectite and sometimes a little chlorite, throughout' (Perrin et al., 1973). The till produces brown, sandy to silty, somewhat tenacious, clay soils. The depth of weathering on the plateaux varies from about 20 cm, usual in the skeletal soils of hill shoulders, to 2 m. The colour changes from unweathered grey to brown at surface, through shades of khaki, with increasing upward decalcification.

The value quoted above for calcium carbonate content refers to the matrix only, but chalk also makes up

over half the pebble-grade and larger erratics. Particularly abundant are subrounded and subangular 2 to 4 mm-diameter pebbles, which locally produce a very chalky gravel within the till. Chalk erratics obtained from fresh sections commonly exhibit striated surfaces.

Flint erratics are also almost ubiquitous within the till. They occur as relatively unworn nodules or as smaller particles down to sand-grade. The nodules are mostly pebble- or cobble-size and may be patinated with varied surface colours, although white is most abundant. Fragmented, commonly angular flints are present, especially as pebble-grade shards, with colours varying from black (the most common) to grey, brown and off-white.

Clasts of other lithologies are much less common and more variable in distribution. Limestone and fine-grained brown sandstone erratics of probable Jurassic age are the most abundant. The calcareous clasts include oyster shells and belemnites. Other erratic lithologies include dark grey to black fissile mudstones (probably Upper Jurassic), septarian nodule cementstones, vein quartz, dolerites, porphyritic microgranites and various gneisses. The large sarsen boulders (silica-cemented sandstone) which occur in the south-west of the district may have been moved by ice, but probably not very far.

Supposed variations of till lithology of regional importance are unconfirmed by the resurvey. One of these was recognised by Harmer (1910), who distinguished between 'chalky-Kimmeridgic' and 'chalky' 'boulder clays' respectively north-east and south-west of a line crossing the north-east corner of the district. Gregory (1898) drew attention to low-lying till on the west side of the Stour valley at Sudbury and suggested that its age differed from that of the neighbouring plateau-top till. Boswell (1929) quoted determinations by Mr Joseph Wright of characteristic Pleistocene foraminifera collected from the former, which Wright said indicated deposition from sea ice. However, the Pleistocene marine microfossils may well be derived from local Crag, and there is no reason to infer a non-Anglian age for this till merely because it lies at a level below that of the till in the surrounding interfluve areas. The line of this valley, like most in the district, almost certainly predates the Anglian glaciation.

Some boreholes in the north-eastern part of the district may have penetrated a till older than the Lowestoft Till, namely the Starston Till of the Suffolk/Norfolk borders (Lawson, 1982), but the evidence is flimsy.

DETAILS

In the north-western and north-central parts of the district, the till sheet is mostly more than 50 m thick below the interfluves. Above the scattered basal sand and gravel bodies, it is largely homogenous, with only rare sand and gravel intercalations away from the main valleys. Some of the largest fresh exposures are along Hartest Brook [8225 5317 to 8272 5273], west of the village; they show up to 1.5 m of bluish grey clay with chalk pebbles, the latter constituting about 15 per cent of the total volume. A typical borehole succession, at Gravelgate Farm, Hundon [7429 4829], proved:

	Thickness m	Depth m
ALLUVIUM	2.6	2.6
TILL		
Clay, brown, becoming grey, with pebbles of chalk, flint and shale	21.4	24.0
Clay, brownish grey, sandy, with mostly coarse, rounded and angular flint pebbles	0.6	24.6

Locally, the macroscopic chalk and/or flint erratic content may increase from the average of 10 to 15 per cent, to more than 30 per cent. Ditches [820 521] near Somerton Hall showed very chalky clay containing abundant scattered, small (less than 2 cm diameter), rounded chalk pebbles and some thin lenses of chalk pebble gravel in a sand matrix.

The till sequences recorded in boreholes in the Cockfield/Brettenham area appear to be more varied than elsewhere. A borehole [9751 5601] near Rattlesden in the north-eastern corner of the district proved (Bristow, 1981):

	Thickness m	Depth m
TILL		
Clay, brown	5.49	5.49
Clay, blue with flints	3.96	9.45
Clay, blue with chalk granules	7.01	16.46
Clay, grey	10.97	27.43
'Boulder Clay and flints'	17.68	45.11
GLACIAL SAND AND GRAVEL	4.57	49.68
CRAG		

Some boreholes in this area showed a brown clay underlying the more widespread bluish grey clay. The former may be equivalent to the Starston Till described by Lawson (1982) from the Suffolk/Norfolk borders, preserved here because of the depression in the Chalk surface mentioned above. For example, a borehole [9223 5544] at Colchester Green Farm, Cockfield, proved:

	Thickness m	Depth m
TILL		
Clay, blue	52.73	52.73
Clay, brown	10.67	63.40
UPPER CHALK		

Farther south, in the Long Melford/Sudbury/Edwardstone area, banding has been observed in the lower part of the till. During the resurvey, the Edwardstone sand and gravel pit exposed, near its eastern end [935 432], about 12 m of till, comprising pale yellowish brown sandy clay, containing apparently randomly distributed flint and chalk pebbles, lying on a flat surface of disturbed Kesgrave Sands and Gravels. However, about 110 m to the north-north-east, comparable pale yellowish brown clay overlies blue-grey till, with scattered small chalk fragments, clearly showing flat-lying multiple colour banding (Plate 6). A section at the northern end of the pit [9334 4344] showed about 6 m of yellowish brown till overlying approximately 4 m of blue-grey till, with subhorizontal 'bedding' defined by several pale, more chalky beds individually up to 0.25 m thick.

At Bear's Pit [8832 4610], near Long Melford, more distinct banding was recorded in the basal 0.5 to 0.6 m of the till (Plate 7). It comprised interlaminated, chocolate-ochreous brown, sandy clays and pale buff to cream, silty clays accentuated in places by yellowish brown sand laminae and stringers of subrounded to rounded flint pebbles with long axes aligned parallel to the bedding. Comparable till in pits at Sudbury, closely associated with Glacial Silt, was described by Boswell (1929, pp.41–43). The stratigraphical position of the banded till, and its association with deposits which were waterlaid, suggests that it originated as flow till and was subsequently overridden by ice from which a more homogenous lodgement till was deposited on it.

In the south-western corner of the district near Ridgewell, the till is quite thick and mostly homogenous, but some differentiation of the lower part is present here too. For example, a well at Birdbrook [7104 4117] proved:

Plate 6   Banded till at Edwardstone gravel pit [9355 4335]. Pale yellowish brown clay till overlying colour-banded blue-grey clay till with scattered small chalk fragments throughout. [A14836].

	Thickness m	Depth m
Dug well	13.7	13.7
TILL		
Clay, hard, with flints and chalk	6.1	19.8
Clay, blue, hard, stony	9.2	29.0
Clay, blue, hard	3.6	32.6
Clay, very hard, stony with chalk in basal 1.6 m	2.5	35.1
Clay, blue, hard	0.9	36.0
Clay, brownish, stony	2.1	38.1
Clay, brownish, sandy	3.0	41.1
'Brown sand, clay, stones and pebbles'	0.8	41.9

UPPER CHALK

The thickness and lithology of the till is more variable nearer the lower Stour valley. North of Gestingthorpe, it is very chalky, and a basal 2.5 to 5 cm-thick calcrete layer occurs locally. The thin and patchy till which caps the high ground near Ballingdon and Bulmer has an irregular base; it fills hollows on the underlying surface. A trial pit [8515 4029] west of Middleton Hall showed:

	Thickness m
Loamy soil	0.3
Clay, greyish brown with ochreous and reddish brown mottle down to 0.8 m; structureless; include flint, vein quartz and brown quartzite; base well defined in some places, gradational in others (interpreted as degraded till)	1.0

KESGRAVE SANDS AND GRAVELS

Another trial pit [8467 4003], near Bulmer, showed 1.6 m of variably sandy, strongly mottled orange, red, brown, grey and pale bluish grey clay till overlying Kesgrave Sands and Gravels.

GLACIAL SAND AND GRAVEL

Under this title are grouped all coarse-grained, partially sorted sediments that are believed to have formed during

Plate 7   Banded till at Bear's Pit, Acton [8832 4610].

The lower part of the till is banded, reflecting sandier layers within a dominantly silty clay deposit.
Rounded flint pebbles and small chalk pellets are present. The basal few centimetres of the deposit, which
overlies Glacial Sand and Gravel, has thin white calcrete bands. The base of the till is indicated by the
hammer head. [A13248]

the Anglian glacial stage. They have been subdivided in
this region by some authors into 'Upper' and 'Lower'
variants according to their relationship to the Anglian till.
In addition, a Barham Sands and Gravels unit, older than
the till and possibly gradational into a 'Barham Soil', the
latter locally developed on Kesgrave Sands and Gravels,
has been identified in some areas. In this district, there is
no consistent difference in the lithological character of
the supposed variants and they are not differentiated.

For the most part, the sand and gravel is extremely
poorly sorted and coarse fractions consist of till deriva-
tives that have suffered minimal additional rounding.
The commonest pebble lithologies are therefore flint
and chalk, each of which may be locally dominant, in
some cases within the confines of a single small mappable body.

The primary distribution of Glacial Sand and Gravel
was restricted both geographically and, with reference to

the penecontemporaneous till, stratigraphically. At various localities, it occurs below, within and above the till
(Figure 18); it is absent from the Anglian sequence in
extensive areas. It seems probable that the bulk of
Glacial Sand and Gravel deposits relates to preferred
north-west to south-east trending tracts, presumably
drainage channels, which have since been either re-
tained or adopted by elements of the present-day river
system. The susceptibility of sand and gravel, compared
with till, to subaerial erosion would favour the latter explanation. These conclusions result from the obvious
concentration of Glacial Sand and Gravel outcrops
along present-day valleys, their apparently inconstant
stratigraphical relationships with till at those outcrops,
and the relative scarcity of sand and gravel in the Anglian sequence of the interfluve till plateaux.

There is one notable exception to the tendency of
major Glacial Sand and Gravel outcrops to occur along

valleys. This is the 0.5 to 1.5 km-wide swathe that runs south and south-east from Acton and Great Waldingfield, through Newton and on to the district margin south of Assington. Borehole data indicate that from Newton northwards this deposit rests on till, but in the Assington area, which is the only part of the outcrop to have been significantly incised by the present river system, sand and gravel locally comprises the whole Anglian stratigraphical sequence. Conjecturally this Acton–Assington swathe may approximate to the general primary relationship of Glacial Sand and Gravel to till: i.e. modern valleys flanked by numerous sand and gravel outcrops may be the sites of formerly continuous thick swathes of Glacial Sand and Gravel.

Thicknesses of Glacial Sand and Gravel are very variable; the following maxima show the scale of deposition: near Clare 11 m (including 1 m of 'Barham' type at the base);

near Stanstead, 11.9 m; near Belchamp Walter, 8.8 m; south of Sudbury, 7.7 m (identified as Barham type); at Great Waldingfield airfield, 6.0 m. An exceptional 21.5 m recorded at Bear's Pit, Acton, is probably related to the tunnel valley which may terminate near this site. All of these figures are quoted from areas covered by IGS Mineral Assessment reports and are therefore reliably distinguished from other clastic deposits by detailed analysis. Areas to the north probably have a similar range of thicknesses, but Glacial Sand and Gravel cannot be reliably distinguished from other sands and gravels in borehole records.

DETAILS

Although Glacial Sand and Gravel is less valuable than the Kesgrave Sands and Gravels as an economic mineral, it has the advantage of ready accessibility and has been extensively

Plate 8   Well-bedded Glacial Sand and Gravel in a pit north-north-east of Glemsford [8146 4964].
Poorly sorted, chalky sand and gravel with low-angle cross-bedding visible at bottom left. Lenses of laminated silt consisting mainly of chalk flour appear as the palest bands. The gravel mostly comprises nodular flint with a variety of other erratics, including Cretaceous and Jurassic sandstones and limestones, and quartz porphyry. [A13243]

worked. Many exposures have therefore been described, although the workings may have since been closed or infilled. The best examples include Bear's Pit [88 46], between Long Melford and Acton (Millward, 1980a), where 4.4 m of till were seen to overlie 10.3 to 18 m of westward-thickening, chalky sand and gravel on Chalk. Much of the gravel showed only traces of bedding and it has been inferred that observed channelling and slumping may be related to the infilling of the buried channel below the Stour valley (see p. 39).

Other notable sections were seen in pits near Stanstead and Glemsford, as follows: at [8495 4934], where a 3.5 m face showed beds tilted 26° towards 69°E; at [8146 4964], where 3 m of poorly sorted but well-bedded chalky sand and gravel are exposed (Plate 8); and at [8208 4923], where 6 m of crudely bedded chalky sand and gravel are interbedded with relatively well-sorted, cross-bedded, chalky gravel and laminated medium- to coarse-grained sand. The pebble assemblage at this latter site is fairly typical for the formation, comprising 80 to 90 per cent of chalk and angular to nodular flint, with minor proportions of limestone, sandstone, siltstone, quartzite, basalt, quartz-feldspar porphyry and *Gryphaea* fragments. Similar ranges of sediments and clast lithologies were recorded from a 4 m gravel-pit face at Smeetham Hall [8415 4110], near Bulmer, and also at Leys Farm [7502 4566], near Clare, where cryoturbation at the top of the 4 m section had caused vertical preferred orientation of pebbles (Millward, 1980b). A recently worked pit [896 559] north-west of Cockfield is in a particularly coarse gravel facies, with some clasts having a greater degree of rounding than is normal.

GLACIAL SILT

The Glacial Silt mapped in the district commonly consists of grey to cream or buff, laminated silt and clay. At some places it can be observed grading or intercalating laterally and/or vertically into other glacigenic sediments. Deposition is inferred to have been in proglacial or subglacial pools characterised by low-energy conditions but with spasmodic inflow of strong currents introducing coarser material. Because of their locally limited extent and close association with other glacial sediments, it has not always been possible to map Glacial Silt separately from the latter. Much of the 'Glacial Silt' of the resurvey was previously referred to as 'Glacial Brickearth', although some of the 'Glacial Brickearth' of the primary survey has been reclassified as Head.

Considerable thicknesses of laminated silts and clays, interbedded in part with till and/or Glacial Sand and Gravel, have also been proved by boreholes along the Stour valley. Woodland (1970) attributed the fine-grained deposits of the East Anglian tunnel-valley infills to a late phase of glaciation, when the subglacial stream, having lost its hydrostatic pressure, became slow-moving or still, facilitating clay and silt deposition. Because of their close association with till and related glacigenic deposits, the laminated clays and silts of both the mapped Glacial Silt and the buried valley infills are assumed to be of Anglian age.

DETAILS

The Glacial Silts were formerly best exposed in the lower Stour valley around Sudbury where they were worked for brickmaking. Some of the pits, now largely degraded or filled, showed diverse folding and faulting of the silts and clays, which were comprehensively described and illustrated by Boswell (1929, pp.40–49, figs. 5, 7–11). His grading analyses of the 'brickearths' below the till emphasised their large silt and clay-grade content, the latter including much calcium carbonate. Most of the structures he described, including complex flow folds, are indicative of formation by slumping and flow of water-lubricated material; they probably originated penecontemporaneously with deposition. However, in one pit at least, step-faulting of planar-bedded 'brickearth' was recorded by Boswell (his figure 8). That locality, 'Gallows Hill', is possibly the same as, or near the pit at a woodmill in east Sudbury [884 419] in which the following section was recorded during the resurvey:

	Thickness m
Till, dark brown, weathered	0.5
Silt and clayey silt, pale brown; laminated in lowest 0.55; pockets of chalky sand	0.9
Gravel, coarse, chalky; angular and nodular flints	1.5
Silt and silty clay, poorly laminated	1.2
Till	

About 5 m of laminated silts and clays with some fine-grained sands were noted in a disused pit [8808 4122] about 650 m east of Sudbury Town Hall. The laminae, which are only a few centimetres thick, show a 25° north-north-easterly dip. The beds are increasingly sandy downwards and in the basal part are contorted into the underlying gravel.

Grey to cream-coloured, chalky silt mapped around Long Melford and Liston is variously associated with till and glacial sand and gravel. It is located above the sides of the buried Stour valley, but 7 m of dark grey silt were also proved within the tunnel-valley infill [8629 4502] below 8.2 m of river terrace gravels at Long Melford. The distribution of laminated silt and clay in the upstream end of the buried channel is indicated in Figure 20. The Clare Borehole proved a total of 53 m of 'silt and clay', interbedded with till and Glacial Sand and Gravel, in the 154 m drift sequence (see p. 40). Smaller patches of Glacial Silt, in close proximity to drift-filled channels, have been mapped at Great Yeldham [758 387] and along the south-west side of the Box valley downstream from Boxford [e.g. 967 391]. Spencer (1967) described the Boxford deposits as buff-coloured, thinly laminated and 'forty to fifty feet thick' (12.2 to 15.2 m).

A large body of Glacial Silt lying under till and on Glacial Sand and Gravel was mapped around Gestingthorpe. It has a mean thickness of about 3 m and was formerly worked for brickmaking. At Wiggery Wood, about 400 m to the east [823 386], laminated silts lie on Woolwich and Reading Beds.

The Glacial Silts mapped in the northern half of the district lie higher in the glacial succession. They include a lens of ochreous brown to pale fawn and brown silt and clayey silt in till at Kiln Farm [862 496] south of Shimpling, and some small bodies of pale grey clay and silty clay within Glacial Sand and Gravel near Cockfield [903 543].

TEN

Post-Anglian drift deposits

Chronological and stratigraphical classifications for the north-west European middle and late Pleistocene are still somewhat unreliable, despite the great advances made in relevant Quaternary research in recent years. The new data include oxygen isotope evidence for worldwide temperature changes and the stratigraphical sequences proved in the North Sea basin, which are much more extensive and continuous than those onshore. During this period, climatic (principally temperature) changes have governed the main variations between marine and terrestrial deposition, and those of sea level, vegetation and fauna. Since the retreat of the Anglian ice, now taken to be about 400 000 years ago (Bowen et al., 1986), several alternations of markedly cold and less cold or even warm climate have occurred. In the past, these have been slotted into a simple glacial/interglacial succession which, with East Anglia and the Midlands used as a type region for the whole of Britain, was classified by Mitchell et al. (1973) as:

Flandrian	Interglacial
Devensian	Glacial
Ipswichian	Interglacial
Wolstonian	Glacial
Hoxnian	Interglacial
Anglian	Glacial

However, this arrangement is now recognised as an oversimplification, in that more than three major cold phases are thought to have occurred. Furthermore, no post-Anglian glaciation is thought to have reached southern East Anglia. Nevertheless, the stages remain as the most convenient stratigraphical classification for the period.

Post-Anglian deposits in this district are largely indicative of former periglacial environments and of changing sea levels. They are discontinous, difficult to date and mostly restricted to the valleys. Periglacial conditions, presumably coinciding with the main phases of glaciation farther north, resulted in solifluxion and deposition of extensive areas of Head, as well as relict features associated with permafrost, such as involutions and frost wedges. Aeolian deposits are known in neighbouring areas but no mappable deposits have been recognised in the Sudbury district.

Changing sea levels have alternately raised and lowered the base-level to which the local streams are graded. Interglacial-phase melting might be expected to coincide with eustatic rise in sea level and the deposition of what we see now as river terraces and associated sediments. Thus, the deposits under the highest terraces of the Stour valley (numbered 3 on the geological map), and the lacustrine silts and clays of Little Cornard, may well be of Hoxnian age, a period generally identified with high (up to 40 m above OD) sea level (Shotton et al., 1977). However, most of the river terrace deposits are probably the products of meltwaters from Glacial-phase snowfields and frozen ground, as envisaged in the Bury St Edmunds district (Bristow, 1990).

LACUSTRINE DEPOSITS

The only Lacustrine Deposits mapped during the resurvey occur near Little Cornard, south of Sudbury. An account of the drift succession exposed in the brick pits there [8885 3843] by Hill (1912) drew attention to 'grey structureless clay' overlying till and containing 'several large masses' of white calcareous material described as '*remanié* Chalk'. The extended, but overgrown sections were cleaned and redescribed by Wilson and Lake (1983), but they found that correlation between the different pits remained problematical (see Figure 23). Hill's 'grey clay' comprises laminated silts and clays; his '*remanié* Chalk' is calcareous tufa intercalated with them. They are underlain by a persistent bed of chalky Glacial Sand and Gravel, and overlain, in the higher part of the pits, by 2 m of flinty Head. The chevron folding and rupturing of the sand and gravel bed near the west end of the pits is ascribed to glacitectonics predating deposition of the silts and clays. The sand and gravel contain reworked Jurassic and Cretaceous foraminifera and ostracods.

The laminated silts and clays are grey and over 4 m thick in the middle pit. In the lower pit, the silt is buff-brown and the clay green-grey. The calcareous tufa in the middle pit is underlain by clayey silts in which the bedding is disturbed, possibly due to the combined effects of channelling and pore water extrusion. The tufa is locally massive and resembles Chalk, as illustrated by Boswell (1929, pl. 2a). It yielded a molluscan fauna identified by Messrs F G Berry and D K Graham of BGS and Dr M P Kerney of Imperial College, as indicating a shallow freshwater habitat, possibly stagnant or slow-flowing, in a cool climate, perhaps comparable with that of the present day. Mr Graham's determinations included *Acroloxus lacustris, Ancylus fluviatilis, Armiger crista, Bathyomphalus contortus, Bithynia tentacula, Lymnaea peregra, Myxas glutinosa, Punctum pygmaeum, Valvata cristata, V. piscinalis, Vertigo substriata, Pisidium henslowanum, P. hibernicum, P. milium, P. nitidum, P. cf. subtruncatum* and *Sphaerium corneum*? The fauna, which is non-age diagnostic, is common in many of the post-Anglian deposits of the neighbouring Bury St Edmunds district (Bristow, 1990, table 6).

HEAD

The term 'Head' is applied to a heterogeneous and lithologically varied group of deposits which characteristically floor valleys and hollows throughout the district; they all postdate the Anglian glaciation. The major agency in

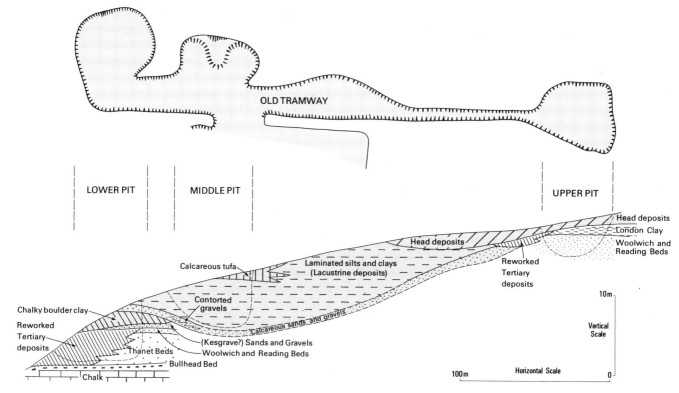

Figure 23 Plan and simplified cross-section of Cornard brick pit (from Wilson and Lake, 1983).

transportation of Head to its present location is assumed to be solifluxion in a periglacial environment, although Head as mapped also includes material emplaced by hillwash and soil creep, which continue under present-day temperate conditions. Thus, the late twentieth-century removal of hedgerows must have accelerated the valley-bottom accumulation of Head.

Head lithologies vary with provenance. The most common parent material in this district is chalky till, although the Head is, presumably, mainly derived from its weathered, more friable and largely decalcified upper part, probably supplemented by wind-blown silt. Consequently, most Head in this district is a rusty brown, silty or sandy, 'loamy' clay with little chalk. A basal flint gravel bed is typical of many Head sections but, where the parent material includes Crag or Glacial Sand and Gravel, the Head may be gravelly throughout. Hence, in the Brad valley between Cockfield and Lavenham, where flint outcrops are common, a considerable proportion of flint gravel is present within the Head. The gravel content (typically flints) may be either generally disseminated, or concentrated in lenses or stringers which, presumably, are either primary waterlaid gravel deposits or lag gravels from which the finer sediments have been winnowed. In the south-east of the district, a very clayey Head is common. This is largely derived from London Clay which is prone to mass movement on all but the gentlest slopes. Recorded thicknesses of Head in the district are as large as 4 m, but 1 to 3 m is more common.

Successive solifluxion lobes, which may indicate accumulation in separate cold periods, have been recognised in the Belchamp Brook and upper Colne valleys (see below). In the upper Glem valley, there appear to have been two distinct phases of Head deposition, one before and one after a major rejuvenation of the main valley. The older Head is in perched pockets, mostly in side valleys at more than 75 m above OD.

DETAILS

Around Long Melford, in addition to the valley Head, there are several expanses of Head occupying hollows in the plateau areas, particularly between the River Glem and Chad Brook. The following section was recorded in a trial pit [8600 4904] 400 m south of Kiln Farm:

	Thickness m
Soil, sandy, clayey, stony	0.3
Clay, greyish brown, silty, sandy in places, with scattered angular and nodular flints up to 10 cm diameter; white patinated flints less than 1 cm diameter; gradational base	0.3
Clay, sandy with coarse angular sand grains up to 1 mm; small pods of ochreous brown clayey sand and abundant angular clasts; sharp irregular base	0.25
Sand, ochreous brown, coarse-grained; some clay; rare scattered angular flints; rapid gradational base	0.15
Clay, pale brown, sandy with grains up to 1 mm; fine to medium gravel lenses, 5 to 10 cm thick; irregular, gradational base	0.15

GLACIAL SAND AND GRAVEL

Farther south, in the Ballingdon area, a trial pit [8623 4050] in a tongue of Head occupying a depression showed rusty brown, medium-grained sand with a lens of dark green, glauconitic clay in the basal 5 cm, and mostly rounded black flints at the base. The deposit thus shows the occurrence of material locally derived from nearby Tertiary strata.

Lower down the Stour valley two types of Head were described by Ellison and Lake (1986, p.52): a common, brown, flinty, sandy clay or silt, and a more local clayey flint gravel, as for example at Kedington Hill [around 890 385]. A little farther west, in the Belchamp Brook valley, a trial pit [8326 4017] showed varied solifucted deposits mostly derived from till:

	Thickness m
Soil	0.3
Sand, dirty yellow to orange brown, fine- to coarse-grained, with pebbles of chalk up to 3 cm across; also flint pebbles	0.2
Clay, medium grey with pale orange mottling increasing towards basal 0.5 m; slightly sandy; scattered pebbles and cobbles including chalk, flint, dark red fine-grained sandstone, grey limestone and dark grey laminated clay; pronounced vertical jointing; irregular patches and lenses, with diffuse boundaries, of silt to fine-grained sand especially near base; sharp but irregular base dipping towards valley at 25° to 30°	1.3–1.8
Clay, grey with ochreous brown patches, slightly sandy, brecciated; chalk granules and rare pebbles; reworked transitional base	0.1–0.2
Silt to fine-grained sand, pale ochreous brown with greyish mottling; variably clayey; abundant chalk cobbles and other erratics including septarian nodules, pale grey limestone, rusty brown fine-grained sandstone, nodular flints and brown quartzite	1.1

In the eastern part of the district, Head deposits, partly covered by alluvium, floor the valleys of the rivers Brett, Brad and Box, and their tributaries. They are especially gravelly in the upper Brad valley and on the side of the Brett valley near Little Farm [9471 5152]. A deep gully [9321 4849] next to Bridge Farm, east of Lavenham, shows 4 m of orange-brown, silty, sandy clay, with chalk and flint pebbles and some crude planar bedding.

RIVER TERRACE DEPOSITS

River terraces cover large areas in the Stour valley, and there are smaller, discontinuous patches in the valleys of Belchamp Brook and the rivers Glem, Brad and Box. The Stour terraces occur at three levels relative to the present valley bottom: the surface of Terrace 1 lies mostly between 1 and 4 m above the adjacent flood plain, Terrace 2 at 5 to 10 m above and Terrace 3 at 10 to 12 m above. The lowest terrace covers extensive tracts along the valley from Wixoe downstream to the southern edge of the district. Terrace 2 is less widespread and has its upstream limit at Stoke by Clare, while Terrace 3 is only recorded downstream from Long Melford.

Terrace 1 merges with the alluvial flat in many places, and there are some terraces with levels intermediate between those of Terraces 1 and 2 and Terraces 2 and 3. Clear-cut distinctions between terrace levels are further limited by the degradation of many terrace surfaces and the restricted size and discontinuous nature of Terrace 3. Three numbered terraces, probably correlatable with some of those described here, were also recognised in the Stour valley of the Braintree district to the south (Ellison and Lake, 1986), but no attempt was made to relate them to the river's longitudinal profile or to terraces in any adjacent valleys. In this district the numbering of Terrace 3, which may be associated with a rather indeterminate knickpoint upstream from Clare, and of Terrace 1, which in longitudinal section roughly parallels the river profile, are both probably chronologically significant. However, the various terraces numbered '2' seem unlikely to represent a single phase of aggradation. The upstream 'Terrace 2', which appears to be tangential with the present-day river profile above Wixoe, possibly predates the downstream Terrace 3.

Most of the terraces are underlain by characteristic river terrace gravel deposits, but lithologies are varied and include fine-grained sediments, probably filling old river channels. It is possible that terrace deposits are thin or absent within some of the areas mapped as River Terrace Deposits. For example, some terrace surfaces in the Stour valley near Glemsford may be erosional features cut into Anglian Glacial Sand and Gravel. In many places, fluviatile terrace deposits are partly or wholly covered by younger solifucted or lacustrine sediments.

Analyses carried out during the sand and gravel resource surveys (Hopson, 1982; Marks, 1982) showed that the Stour terrace deposits of the Sudbury area consist of cross-bedded flint or sandy gravels with subordinate silty sand and silt seams. Locally, they have pyritous, carbonaceous, dark greenish black, sandy silts at surface which may be the younger infill of abandoned distributary channels. Between Cavendish and Wixoe the terrace deposits of the Stour valley are characteristically very clayey, sandy gravel largely composed of angular patinated flint with a smaller proportion of chalk, generally with a thin cover of pebbly, sandy silt.

Thicknesses of the Stour terrace deposits are variable. Boreholes drilled into First Terrace deposits at both Long Melford [8629 4502] and Great Cornard [8862 4022] proved 8.2 m of sand and gravel overlying glacial deposits. Conversely, river terrace deposits on some bench features, such as one of Terrace 2 [around 861 461] west of Long Melford, are thin and patchy. More typical thicknesses vary from 2.5 to 5 m.

All the terrace deposits postdate the Anglian glaciation; Terrace 2 gravels at Brundon near Sudbury have yielded artefacts and mammalian bones of probable late Wolstonian to Ipswichian age (Wymer, 1985 and see below). Some of the higher terraces in the lower Stour valley may be related to the perched Head deposits in the upper Glem valley which would indicate a cold climate origin, probably during the Wolstonian.

DETAILS

The deposits of a terrace intermediate in height between Terraces 1 and 2, between Cavendish and Glemsford, have been worked for brickmaking. In a trial pit [8233 4683], the following section was recorded:

	Thickness m
Soil	0.2
Silt, brown; grading down into orange-brown, clayey and silty flint gravel and coarse-grained sand	2.3
Clay, khaki-brown, very silty and sandy, with abundant chalk pebbles and flints	1.0
Silt, laminated with pale grey and brownish grey bands; scattered chalk granules; some brown mud laminae up to 5 cm thick near base; lenses, stringers and pockets of fine flint gravel; shelly pockets yielding a fauna dominated by *Pupilla muscorum*	0.5

Mr D K Graham comments that the fauna is compatible with a cold, marshy environment.

The disused pit [863 417] south of Brundon was dug to extract Terrace 2 gravels from below Head deposits. The following section was recorded in the west face:

	Thickness m
HEAD	0.8–1.2
TERRACE DEPOSITS	
Sand and Gravel; gravel predominantly angular flints but also includes rounded and nodular flints as well as brown sandstone, quartzite and vein quartz pebbles; matrix is mostly ochreous to rusty brown sand; pebbles imbricated; cryoturbated	0.8–1.5
Sand and Gravel; content much as above; well-bedded with some beds of clean yellowish brown sand	1.4

Cleaner material yielding mammalian bones was revealed by excavation in the northern part of the pit, but it was not possible to relate any exposures seen during the resurvey with the sequence recorded here by Moir and Hopwood (1939). They described mammalian remains including horse (very common), bison (common), auroch (common), wolf, bear, lion, rhinoceras, red deer, Irish (Giant) deer, mammoth (common) and elephant, as well as thirty species of freshwater and terrestrial molluscs, collected from the base of terrace gravels up to 4.6 m thick. Both molluscs and mammals were ascribed to a grassland environment — the mammal fauna suggesting a drier and slightly warmer climate than the present one. Bone samples from the site were given a Uranium series data of 174 000 ± 30 000 by Szabo and Collins (1975). The gravels also yielded Palaeolithic flint artefacts, including axes and 'tortoise' cores. Although clearly derived, Wymer (1985) considered that they mainly represent one 'industry', which he identified as Levalloisian, indicating a late Wolstonian to Ipswichian age.

The large embayment on the east side of the Stour between Great and Little Cornard is largely floored by Terrace 1 gravels of some thickness (see p. 51). The terrace is mostly overlain by a thin veneer of sand, but part of an abandoned river channel next to the valley side, known as Cornard Mere, contains peat.

River terrace sand and gravel deposits at levels related to the First and Second terraces of the Stour valley also occur in the lowest 4 km of the Glem valley. Gravel has been extracted from below the large, poorly defined terrace [around 8365 4880] north-east of Glemsford. The flat upon which Hartest stands is incised by the brook running through the village. It is underlain by up to 3.5 m of brown silty clay with gravel lenses and may correlate with one of the higher Stour valley terraces.

ALLUVIUM

Alluvium occurs along the valleys of all the major streams. It underlies flood plains with widths varying from less than 20 m, to as much as 600 m in the lower Stour valley. The characteristic near-surface lithology is a pale to medium brown, silty clay or silt, locally with a high organic content and grading to peat. Many of the stream bank sections cut into alluvial flats also show flint gravel — either as lenses or stringers within the silt and silty clay or as apparently more continuous gravel beds underlying the argillaceous alluvium. The latter gravels may equate in age with valleyside terrace deposits but at no place are there sufficient boreholes to prove contiguity. Comparable gravels below river floodplains occur in adjacent districts (Ellison and Lake, 1986, p.49; Bristow, 1990, p.77). Because existing stream beds are commonly floored by gravels, it is inferred that most of the finer-grained alluvium comprises overbank sediments. In some of the wider valleys, notably that south of Sudbury, the channels of abandoned meanders are infilled with peat.

In the Stour valley south of Sudbury, alluvial silt has been proved to a depth of 9.2 m, but alluvium thicknesses of 2 to 5 m are probably more typical. Although some alluvial sequences are shelly and others overlie peat beds, no indication of age has been obtained. However, the alluvium probably ranges in age from late Devensian, up to the present day.

DETAILS

The Stour valley is wide and badly drained near Water Hall, Wixoe. A borehole [7085 4497] near Cotton Hall proved:

	Thickness m	Depth m
Soil	0.4	0.4
Clay, brown, silty	0.9	1.3
Silt, bluish grey, with peat and small chalk pebbles	1.2	2.5
Peat, silty, with molluscan remains; seams of black angular flints	1.6	4.1
Gravel, with some silt and sand	9.1	13.2

UPPER CHALK

A borehole [7915 4526] 2 km downsteam from Clare proved 4 m of brown alluvial clay with shell debris, peaty seams and, near its base, a thin gravel bed, all overlying chalky till. The best exposures of alluvium in the Glem valley are probably river bank sections between Boxted and Somerton Hall [about 824 514] showing up to 1.5 m of brown silty clay. The floodplain of Chad Brook, in its lowest part downstream from

Bridge Street, is up to 130 m wide and overlies soft, brown and grey silty clay.

Between Long Melford and Sudbury, the alluvium is mostly about 1.5 to 2 m thick and consists of soft, dark brown, silty clay, commonly with a high percentage of organic matter, overlying flint gravel. A borehole [8582 4497] near Bridge House, Liston, proved 3 m of silty clay overlying 6.2 m of gravel, on Chalk. South of Sudbury, the alluvium is thicker, as shown by two boreholes; one at Shalford Meadow [8822 3865] proved:

	Thickness m	Depth m
Silt, olive-grey, clayey	0.7	0.7
Silt, grey and brown, clayey; shell debris and peat	1.3	2.0
Silt, greenish blue, with wood fragments	5.0	7.0
Silt, bluish grey, sandy, with some pebbles	2.2	9.2
UPPER CHALK		

and another, at Lower Farm [8857 3775] revealed:

	Thickness m	Depth m
Soil	0.2	0.2
Silt, mottled brown and bluish grey, clayey, with shell debris, peat and fragments of wood	5.2	5.4
Gravel	5.2	10.6
UPPER CHALK		

The other major areas of alluvium are along the valleys of the rivers Brett, Brad and Box in the eastern part of the district, and along Belchamp Brook and the upper Colne in the southwest. The flood plain of the Colne, and that of its tributary north-west of Great Yeldham, are both narrow and underlain by silts and loams little more than one metre thick. The alluvial strip along Belchamp Brook is almost 300 m wide near Belchamp Walter church. Ditches show exposures of soft, dark brown, peaty clay or very silty clay. There are exposures of brown, silty alluvial clay as much as 1.5 to 3 m thick in both the Brad and upper Brett valleys, but total alluvium thicknesses are unlikely to be much greater. There is a wide alluvial flat at the confluence of the two rivers underlain by grey-brown silty clay,

humic in part. The alluvium of the upper Box valley is mostly a peaty silty clay which grades into peat on the sides of the flood-plain at several places [e.g. 940 410] between Edwardstone and Boxford.

PEAT

The surface peat of this district is recent in origin and is of the 'fen' or 'sedge' type developed on permanently wet low-lying land surfaces. The mapped occurrences lie in the south-east of the district in valley bottoms westwards and northwards from Boxford [e.g. 943 408 and 967 413]. There, the necessary constant source of water is mostly a spring line at the top of the London Clay, which is overlain by Kesgrave Sands and Gravels with Glacial Sand and Gravel above. However, mappable peat (i.e. more than a metre thick) also occurs farther up the Box valley, between Edwardstone and Great Waldingfield [e.g. around 935 429], north of the London Clay outcrop. The valley-bottom alluvium throughout the district commonly contains at least a proportion of organic material, albeit not enough to classify the sediment as peat.

LANDSLIP

The only mapped landslip areas are on and just above the London Clay outcrops in the south-east of the district. Slopes there are liable to instability due to the spring water which commonly flows from the junction between the London Clay and overlying Kesgrave Sands and Gravels. The resulting high moisture content and pore pressures in the London Clay reduces its strength and causes any slope of more than 7° or so to be potentially unstable. The spring-water lubrication of the junction itself is also likely to lead to landslip of the overlying deposits. One such area of landslip occurs on the valley side north of Boxford [around 966 415], where the slope has collapsed in a series of small rotational or translational failures.

ELEVEN

Economic geology

CHALK

There are no working chalk quarries in the district although chalk has been extracted in the past from many small to medium-sized pits in and around Sudbury, as well as from small quarries at Bulmer [8321 4006], Kedington [7095 4612] and Monks Eleigh [9662 4717] among others. Some of the Sudbury pits were operating at the time of the primary geological survey in the 1870s, but even then many were already disused. Perhaps most of the workings were multipurpose, with overlying Tertiary and/or Quaternary brick clay or sand and gravel also being utilised. At least two of the Sudbury area pits, at Great Cornard, had extended underground workings (see p.59).

The main use of the chalk was probably agricultural, either as screened chalk or burnt lime. Despite the local conspicuousness of flint buildings (see Plate 9), it is unlikely that the extraction of flints was a principal objective of the chalk workings, as the Upper Chalk around Sudbury is particularly deficient in flint bands and, furthermore, loose flints can be obtained easily from the local till.

Although the subdrift Chalk outcrop covers half the district, it cannot be considered as representing an important short- or medium-term resource. In the plateau areas, the till overburden is 20 to 60 m thick, and most of the drift-free valleyside Chalk outcrops are in environmentally sensitive areas. A further economic limitation is that, although the softness of the Chalk in south-east

Plate 9 The parish church at Long Melford.
Dressed stone from outside the district has been used for the quoins and construction of the delicate tracery at the east end. Squared flints, the only natural building stone in the district, infill the bulk of the walls. [A13251]

England permits its extraction by mechanical diggers without blasting, the high water content of the porous rock commonly restricts small operations to summer-only production.

SAND AND GRAVEL

As part of the national assessment of naturally occurring aggregate, the Industrial Minerals Assessment Unit (IMAU) of the Institute of Geological Sciences (now the British Geological Survey) carried out a number of surveys in this district for the Department of the Environment. Six surveys were completed during the period 1975–81 and the reports were published in 1981 and 1982 (see Figure 24). Together they cover some 277 km² (i.e. 50 per cent approximately of the Sudbury sheet). It is from these reports that much of the information concerning potentially workable deposits in the Crag, Kesgrave Sands and Gravels, Glacial Sand and Gravel and River Terrace Deposits is derived. A summary of the results from these surveys, together with archive borehole data, is shown in Figure 25. A list of major sand and gravel operations is given in Table 3.

For the purpose of the national survey of aggregate resources, a deposit was regarded as potentially workable if it met the following four arbitrary criteria:

1. The deposit should average at least one metre in thickness.
2. The ratio of overburden to sand and gravel should not exceed 3:1.
3. The proportion of fines (particles less than 0.063 mm diameter) should not exceed 40 per cent (see Figure 26).
4. The deposit should be within 25 m of the surface.

Crag

The Crag in this district is generally dark reddish brown due to weathering and grades as sand to pebbly sand, with rare sandy gravel at the base. Towards the north-east of the sheet, at depth, the Crag becomes greyish green and contains discrete thin clay/silt seams and laminae which impart an overall 'clayey' nature to the deposit. Shell debris is rare within the weathered material.

Much of the gravel is concentrated in a basal lag of varying thickness and is composed principally of well-

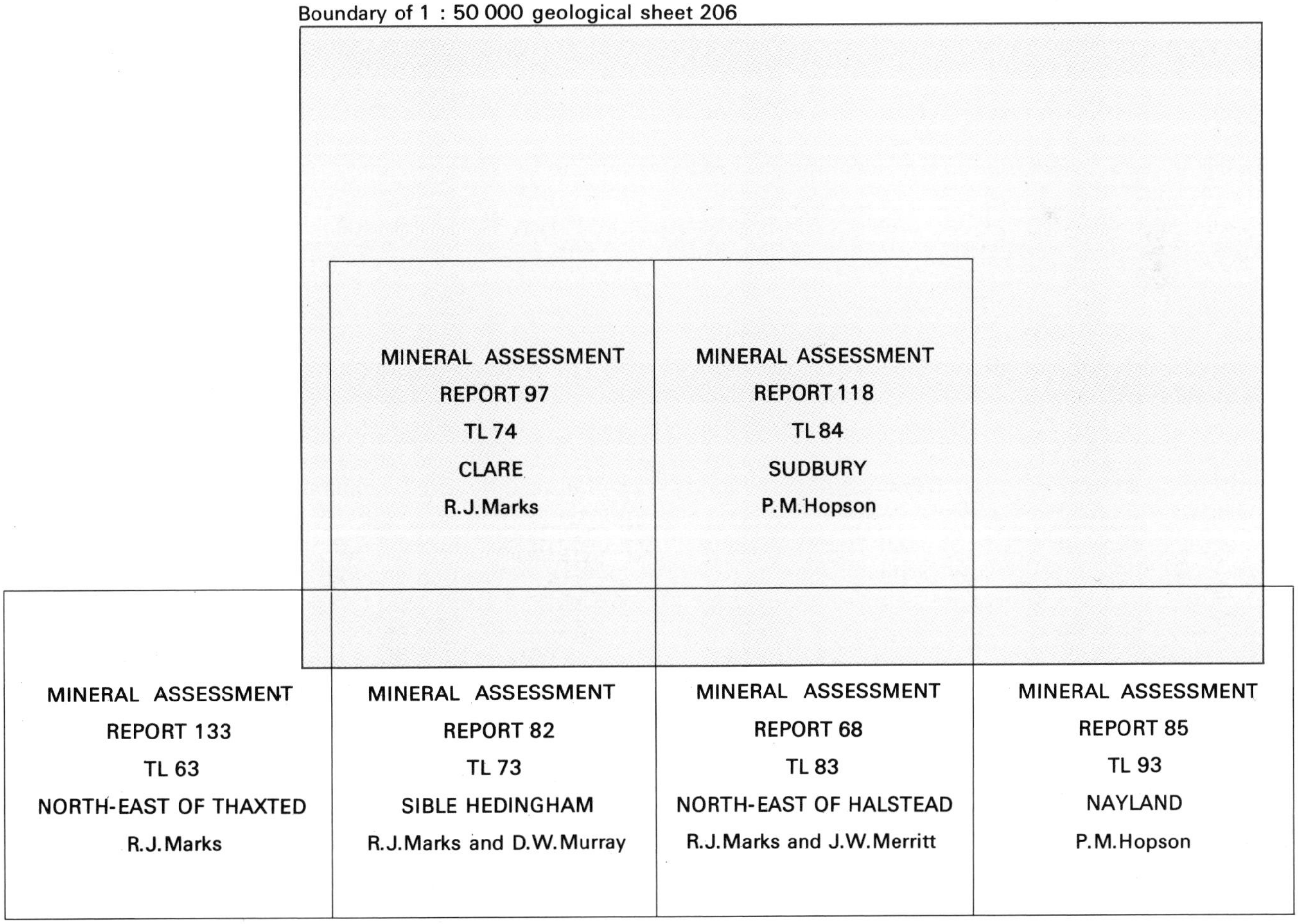

Boundary of 1 : 50 000 geological sheet 206

| MINERAL ASSESSMENT REPORT 97 TL 74 CLARE R.J. Marks | MINERAL ASSESSMENT REPORT 118 TL 84 SUDBURY P.M. Hopson |

| MINERAL ASSESSMENT REPORT 133 TL 63 NORTH-EAST OF THAXTED R.J. Marks | MINERAL ASSESSMENT REPORT 82 TL 73 SIBLE HEDINGHAM R.J. Marks and D.W. Murray | MINERAL ASSESSMENT REPORT 68 TL 83 NORTH-EAST OF HALSTEAD R.J. Marks and J.W. Merritt | MINERAL ASSESSMENT REPORT 85 TL 93 NAYLAND P.M. Hopson |

Figure 24 Plan showing geographical cover of Mineral Assessment Reports in the district.

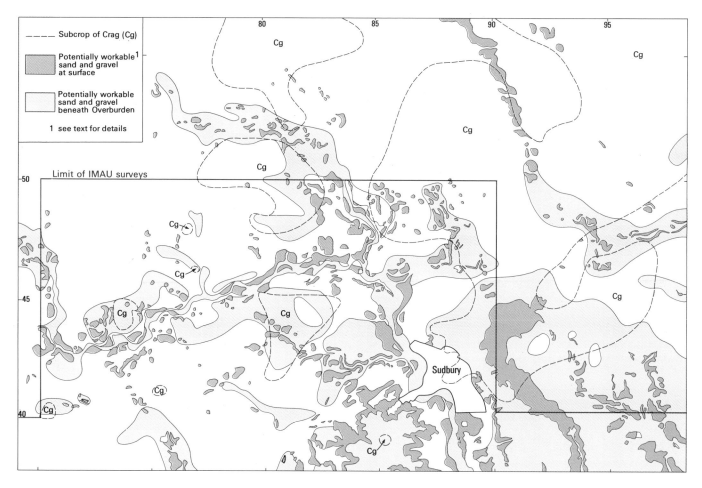

Figure 25 Summary assessment of sand- and gravel-bearing deposits in the district.

rounded, black flints and dark brown, rounded, phosphatic nodules. Higher deposits include iron-rich, partially cemented, sand seams. Overall, the gravel fraction is composed of subequal proportions of angular, subangular and well-rounded flint, with some quartz, quartzite and rare sandstone.

The sand is predominantly medium and fine grained with some of coarse grade and shows bimodal distribution across the grades. Generally, the sand is subrounded to rounded quartz, with some angular to subangular flint, and traces of rounded phosphatic pellets, tabular shell debris and tabular ironstone. Because of the included phosphate, ironstone and overall iron-staining on most grains, the sand is unsuitable as a mortar sand.

Kesgrave Sands and Gravels

This formation forms the greatest aggregate resource in the district and can be divided into sand and pebbly sand/sandy gravel units. The sands predominate to the north and west, while the pebbly unit has its greatest development in the south and east. Commonly, the sands have discrete clay laminae within them which impart a general 'clayey' or 'very clayey' nature to the deposit.

Overall, the gravel fraction is composed of angular and subangular flint, with well-rounded flint, quartz and quartzite in subequal proportions, together with rare sandstone, chert and igneous material.

The sand fraction within the sandy unit of the formation is strongly unimodal in distribution, being a single grade of either fine-grained or medium-grained sand. It is generally composed of subangular to subrounded quartz, but also contains appreciable quantities of mica in places, particularly in the clay/silt laminae. Within the pebbly unit, the distribution of the sand fraction is broadly unimodal, becoming bimodal towards the base where reworking of underlying Crag sands has occurred. It grades as predominantly medium grained with subordinate fine and coarse grained, and is composed mainly of subangular, with some rounded quartz and subangular flint. The latter becomes predominant in the coarse sand fraction.

Barham Sands and Gravels

The Barham Sands and Gravels were identified in many boreholes, although mapping of the unit was found to be impracticable. Generally, they are too 'clayey' for beneficiation and are used in an 'as dug' condition for fill, or discarded as waste. They contain chalk in places,

Table 3 Major sand and gravel operations.

	Grid references	Geological deposit worked
Working pits		
Edwardstone	933 434	Kesgrave Sands and Gravels
Cockfield	896 557	Glacial Sand and Gravel
Disused pits		
Great Yeldham	755 390	Glacial Sand and Gravel
Great Cornard 1	887 395	First Terrace
Great Cornard 2	887 391	First Terrace
Great Cornard 3	885 389	First Terrace
Hagmore Green	963 390	Kesgrave Sands and Gravels
Stoke Row	763 444	Undifferentiated Terrace Deposits
Wixoe	709 432	Undifferentiated Terrace Deposits
Chilton Street	746 468	Glacial Sand and Gravel
Brundon west	862 417	Second Terrace
Brundon north	863 418	Second Terrace
Newton Road	894 417	Barham/Kesgrave Sands and Gravels
Acton	884 463	Glacial Sand and Gravel

showing an intermediate character between the Kesgrave Sands and Gravels and the overlying chalky till.

Glacial Sand and Gravel

This is the most difficult aggregate resource to quantify, being the most variable in stratigraphical position, grade, composition and thickness. It is generally associated with the chalky till, being found below, above, and in lenticular masses within it. It has also been proven at considerable depth within the buried valley of the River Stour west of Long Melford.

The gravel fraction is composed predominantly of angular flint, with subequal proportions of subordinate well-rounded flint, chalk, quartz and quartzite, and rare sandstone, igneous and metamorphic rocks, and fossil debris (principally Jurassic). Within the buried valley sequences, most notably that of the Stour valley, sands almost entirely composed of chalk are found associated with laminated glacial silts.

Throughout the deposit, sand is generally of medium and coarse grade, with subordinate fine grade. It is composed of angular flint and subequal proportions of chalk and subangular quartz. The finer grades become increasingly quartz-rich.

River Terrace Deposits

River Terrace Deposits grade overall as 'clayey' sandy gravels, but range from open framework gravels to clay-bound pebbly sands. Some of the lower River Terrace Deposits are beneath the water table, whereas higher terraces tend to have thin perched water tables at their base.

The gravel fraction is generally fine, with subordinate coarse gravel. It is composed overwhelmingly of angular

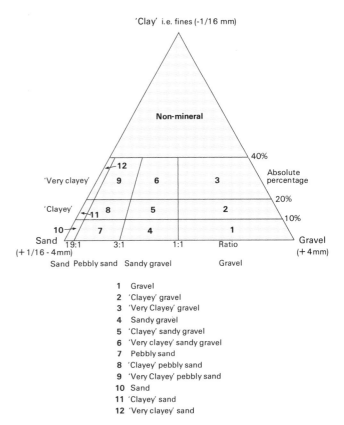

1   Gravel
2   'Clayey' gravel
3   'Very Clayey' gravel
4   Sandy gravel
5   'Clayey' sandy gravel
6   'Very clayey' sandy gravel
7   Pebbly sand
8   'Clayey' pebbly sand
9   'Very Clayey' pebbly sand
10 Sand
11 'Clayey' sand
12 'Very clayey' sand

Figure 26 Diagram to show the descriptive categories used in the classification of sand and gravel.

flint but contains pebbles of quartz, quartzite, chalk, sandstone and rare limestone and igneous material. The proportion of quartz rises significantly in the finest gravel fraction.

The sand is medium to coarse grained, with some fine material. It is composed of predominantly angular flint and subangular quartz, with traces of chalk, ironstone and sandstone.

Fines are usually in the form of disseminated silt-grade material, but clay enrichment can occur at the surface above the water table.

BRICK CLAY

The only current brick clay working in the district is in London Clay at 'The Brickfields' (Bulmer brick pit) [833 382] between Bulmer Tye and Gestingthorpe. Here, silt and clayey silt is extracted from a 7.8 m sequence of interbedded silts, sands and cementstones to produce facing bricks (see Plates 4 and 5). Disused brick clay pits abound elsewhere in the district. The deposits previously worked include 'Lower London Tertiaries' at Gestingthorpe [around 823 395] and Ballingdon Grove [871 403], Glacial Silt at Gestingthorpe [818 387], Boxford [around 965 405] and the 'Alexandra' [around 884 419] and 'California' [883 422] brickworks in Sudbury, till near Hawkedon [787 537] and probable Head at

Great Yeldham [around 380 758]. Most of these localities were described by Boswell (1929). The most recently closed (about 1960) and best-documented of the disused pits is at Little Cornard [8885 3843]. This pit worked a complex suite of Tertiary and Quaternary deposits (see Figure 23) for the manufacture of handmade bricks: white ones from interglacial lacustrine silts and clays, supplemented with Tertiary sands from higher in the pit, and red ones from the partly reworked 'Lower London Tertiaries' near the bottom of the pit.

HYDROGEOLOGY AND WATER SUPPLY

The district is contained within the River Stour catchment except for parts of the extreme north and the south-west corner. The headwaters of the Stour receive water via the Ely–Ouse to Essex Transfer Scheme, which is reabstracted at Wixoe for transfer to the rivers Colne or Pant/Blackwater. The Stour Augmentation Groundwater Scheme allows nine boreholes to pump 30M m³ of water annually from the Chalk to the river, to provide a net gain to river flow of more than 60 per cent. This water is subsequently reabstracted downstream.

Annual average rainfall is 600 mm and annual protection evaporation is about 535 mm; persistent summer soil-moisture deficits are normal. The main aquifer is the Chalk, which underlies the entire district but which is largely concealed by till and locally confined. Small volumes of permeable material occur in the Palaeocene deposits, the Crag and younger Quaternary sands and gravels.

The hydrogeology of the Sudbury district is summarised in the Institute of Geological Sciences 1:125 000-scale Hydrogeological Map of Southern East Anglia published in 1981. Groundwater potential in the district was reviewed by Whitaker (1906), Whitaker and Thresh (1916), Woodland (1946) and Cole et al. (1965). More recent work is described in the reports of the former Anglian Water Authority Essex River Division

Chalk

The uppermost 10 to 20 m of Chalk below its eroded surface, particularly where it is marly and weathered, may be putty-like and poorly jointed. Normal joint patterns are more apparent in fresher material where the conventional groundwater hydraulics of the Chalk prevail. Groundwater is stored partly within the intergranular matrix of the rock and partly in the microfissure system, but transport of groundwater relies on conduit flow through fissures, one millimetre or so in width, and in the smaller microfissures. Fissuring tends to be greatest in lower-lying areas where stress release from diminished overburden and enlargement of resultant fissures by solution offer the most favourable locations for production boreholes.

Buried channels, in particular that below the Stour valley, inhibit groundwater flow in the Chalk and secondary permeability may be weak in the vicinity as a result.

The potentiometric surface of the groundwater (the imaginary surface that represents its static head) tends to a subdued form of the surface topography. It attains a maximum elevation of over 70 m above OD in the areas around Rede and around Cowlinge at the topographical divide which marks the northern limit of the Stour catchment. Although both areas are covered in till, their relatively high groundwater level demonstrates that recharge takes place through this weakly permeable cover. The potentiometric surface declines at an average gradient of 1 in 300 towards the east and south, and is at only 20 m above OD beneath the River Stour at Great Cornard. Groundwater flow occurs in the direction of this prevailing hydraulic gradient and locally also towards the valley bottoms.

The hydraulic properties of the Middle and Upper Chalk are essentially similar and together they form a single hydrogeological unit. Most variation occurs between valley and upland sites, and with the nature and thickness of the cover (compare the properties of the Hundon Borehole, which is situated at an elevation of 75m above OD, with the Stour valley boreholes at Wixoe and Sudbury in Table 4). Transmissivity (the rate at which groundwater is transmitted through a unit width of aquifer) ranges from only 50 m²/d (square metres per day) at Hundon to 1500 m²/d at Wixoe. Storativity (the volume of water given up per unit horizontal area of the aquifer) lies in the range 5×10^{-4} to 2×10^{-2}. Borehole performance is best where the Chalk is at or

Table 4 Results of pumping tests on the Chalk aquifer.

Location	National Grid reference	Open Hole Contact with Chalk (m.b.g.l.)	Static Water Level (m.b.g.l.)	Sustainable Yield (l/s)	Drawdown (m)	Transmissivity (m²/d) (see above)	Storativity (see above)
Great Wratting	691 481	30–91	2	5	8	400	2×10^{-2}
Wixoe	710 432	33–91	5	121	12	1500	5×10^{-3}
Sudbury (Little Cornard)	889 392	32–91	1	57	7	1000	2.5×10^{-3}
Hundon	733 487	45–107	19	9	32	50	5×10^{-4}

m.b.g.l. = metres below ground level

Table 5 Typical chemical analyses of borehole waters from the Chalk in milligrams per litre.

Location	Brettenham [9714 5408]	*Ashen House Farm [7484 4274]	Claredown Farm [7778 4472]	*Stafford Works Long Melford [8471 4572]	*Guilford Kapwood Ltd Great Cornard [8860 4020]	*Boyton Hall [973 465]	Stone Street Boxford [9669 3965]	*Hill Farm Boxford [9636 3846]
Date of analysis	21/5/35	17/4/89	11/1/80	17/4/89	17/4/89	17/4/89	29/3/77	17/4/89
pH	7.2	7.2	6.8	7.1	7.2	7.5	7.4	7.1
Electrical conductivity (μS/cc)	–	986	890	815	711	882	880	815
Total dissolved solids	860	877	550	675	594	720	606	675
Bicarbonate (HCO_3^-)	171	403	354	387	353	381	439	387
Sulphate (SO_4^{2-})	192	207	75	63	22	66	43	63
Chloride (Cl^-)	184	44	32	49	49	80	94	49
Nitrate (NO_3-N^-)	0	0	1.1	0	4.8	0	<0.2	0
Calcium (Ca^{2+})	123	158	130	120	125	120	126	120
Magnesium (Mg^{2+})	44	36	11	16	5	14	34	16
Sodium (Na^+)	119	22	17	34	15	53	50	34
Potassium (K^+)	–	5	5	4	3	6	5	4
Iron (total)	1.20	2.72	<0.05	0.89	0	0	0.06	0.89
Fluoride	–	0.25	0.25	0.60	0.25	0.98	0.06	0.60

* Data provided by the Anglian Region (Eastern Area) of the National Rivers Authority.

μS/cc (= microsiemens per cubic centimetre)

near outcrop. Borehole yield is partly dependent on borehole diameter; the larger the diameter, the greater the hydraulic contact with available fissures. Thus, the large borehole at Wixoe, with a completed diameter of 610 mm, has a high yield of 121 l/s (litres per second) (Table 4). Typically, boreholes are completed to 150, 200 or 250 mm diameter, penetrating up to 50 m of saturated Chalk, to provide modest sustainable yields of 1 to 5 l/s with corresponding drawdowns up to 30 m.

Annual groundwater levels fluctuate up to 2 m. Typically, the potentiometric surface is lowest during October or November, and highest some time between February and May.

The Chalk groundwater is of the calcium bicarbonate type with total dissolved solids in the range 600 to 900 mg/l (milligrams per litre) (Table 5). Nitrate concentrations are acceptably low in the district. The chloride and sulphate ions attain highest concentrations where the till is thickest, e.g. 184 mg/l and 192 mg/l respectively at Brettenham; the sulphate derives from finely disseminated gypsum within the till. Pore water in till at Hundon increases from 100 mg/l near surface to over 1100 mg/l at 29 m depth, decreasing to 500 mg/l at 30.5 m, 1 m above the underlying sands and gravels. A ^{14}C date of 14 000 years BP for Hundon Chalk water suggests that recharge to the intervalley areas may be small.

'Lower London Tertiaries'

The hydraulic properties of the 'Lower London Tertiaries' are largely unknown. These strata are normally cased out and boreholes continued down into the Chalk because the water in the 'Lower London Tertiaries' is largely ferruginous. Most of the strata are in any case unsaturated but there are some domestic supplies from shallow wells.

Crag and younger Quaternary sands and gravels

The Crag and the younger Quaternary sands and gravels are hydraulically similar and commonly represent a single hydrogeological unit. A number of modest-yielding boreholes and wells are licensed to abstract from these strata, principally for spray irrigation, but also for other agricultural and industrial purposes. There are also a number of domestic supplies from boreholes, the sustainable yields of which rarely exceed 1 l/s; water quality is generally poor. Groundwater in these sands and gravels, especially in the Crag, tends to be depleted in oxygen, and iron may be present in solution.

GROUND INSTABILITY

During the 1970s and early 1980s some recently constructed buildings in the northern part of Great Cornard were affected by ground collapse into old underground chalk workings (Pearman, 1982, 1983). These mines were found to consist of tunnels up to 38 m long, extending generally south-eastwards from the floor of a series of open chalk quarries between Newton Road [around 887 414] and King's Hill [around 883 409]. The two separate areas affected (Maldon Court and Pot Kiln School) were investigated by drilling and geophysical surveys, and the cavities were delineated and filled in. However, given the abundance of old chalk workings in and around Sudbury and their nearness to much new construction, the possibility of further comparable subsidence problems cannot be discounted.

Landslips on and above the London Clay outcrops in the south-eastern parts of the district are discussed on p. 53.

REFERENCES

Most of the references listed below are held in the Library of the British Geological Survey at Keyworth, Nottingham. Copies of the references can be purchased subject to the current copyright legislation.

AGUIRRE, E, and PASINI, G. 1985. The Pliocene–Pleistocene boundary. *Episodes*, Vol. 8, 116–120.

ALLEN, P (Editor). 1984. *Field guide (revised edition) to the Gipping and Waveney valleys, Suffolk. May, 1982.* (Cambridge: Quaternary Research Association.)

ALLSOP, J M. 1985. Geophysical investigations into the extent of the Devonian rocks beneath East Anglia. *Proceedings of the Geologists' Association*, Vol. 96, 371–379.

— and SMITH, N J P. 1988. The deep geology of Essex. *Proceedings of the Geologists' Association*, Vol. 99, 249–260.

BADEN-POWELL, D F W. 1948. The Chalky boulder clays of Norfolk and Suffolk. *Geological Magazine*, Vol. 85, 279–286.

BAILEY, H W, GALE, A S, MORTIMORE, R N, SWIECICKI, A, and WOOD, C J. 1983. The Coniacian–Maastrichtian stages of the United Kingdom, with particular reference to southern England. *Newsletters on Stratigraphy*, Vol. 12, 29–42.

BARKER, R D, and HARKER, D. 1984. The location of the Stour buried tunnel-valley using geophysical techniques. *Quarterly Journal of Engineering Geology*, Vol. 17, 103–115.

BENNETT, F J, and BLAKE, J H. 1886. The geology of the country between and south of Bury St Edmunds and Newmarket. *Memoir of the Geological Survey of Great Britain.*

BERGGREN, W A, KENT, D V, and FLYNN, J J. 1985. Palaeogene geochronology and chronostratigraphy. 141–195 *in* Geochronology of the geological record. SNELLING, N J (editor). *Memoir of the Geological Society of London*, No. 10.

BERRIDGE, N G. 1989. Geological notes and local details for 1:10 000 sheets TL94NW and SW: Lavenham and Great Waldingfield. *British Geological Survey Technical Report*, WA/89/46.

BOSWELL, P G H. 1916. The stratigraphy and petrology of the lower Eocene deposits of the north-eastern part of the London Basin. *Quarterly Journal of the Geological Society of London*, Vol. 71, 536–591.

— 1929. The geology of the country around Sudbury (Suffolk). *Memoir of the Geological Survey of Great Britain.*

— 1952. The Pliocene–Pleistocene boundary in the east of England. *Proceedings of the Geologists' Association*, Vol. 63, 301–312.

BOWEN, D Q, RICHMOND, G M, FULLERTON, G S, SIBRAVA, V, FULTON, R J, and VELIOCHKO, A. 1986. Correlation of Quaternary glaciations in the Northern Hemisphere. 509–510 in Quaternary glaciations in the Northern Hemisphere. SIBRAVA, V, BOWEN, D Q, and RICHMOND, G M (editors). *Quaternary Science Reviews*, Vol. 5.

BREISTOFFER, M. 1940. Revision des Ammonites du Vraconien de Salazac (Gard) et considerations generales sur ce sous-étage albien. *Travaux du Laboratoire de Géologie de la Faculté des Sciences de Grenoble*, Vol. 22, 1–101.

— 1947. Sur l'age exact des grès verts de Cambridge (Angleterre). *Compte Rendu Sommaire des Séances de la Société Géologique de France*, Vol. 15, 309–312.

BRISTOW, C R. 1980. The geology of the country around Walsham le Willows: Explanation of 1:10 560 geological sheet TM07SW. *Open-File Report of the Institute of Geological Sciences*, No. 1980/6.

— 1981. Geology of the country around Felsham and Rattlesden. Explanation of 1:10 560 geological sheets TL95NW and 95NE. *Open-File Report of the Institute of Geological Sciences*, No. 1981/1.

— 1983. The stratigraphy and structure of the Crag of mid-Suffolk, England. *Proceedings of the Geologists' Association*, Vol. 94, 1–12.

— 1985. Geology of the country around Chelmsford. *Memoir of the British Geological Survey*, Sheet 241 (England and Wales).

— 1990. Geology of the country around Bury St Edmunds. *Memoir of the British Geological Survey*, Sheet 189 (England and Wales).

— and COX, F C. 1973. The Gipping Till: a reappraisal of East Anglian glacial stratigraphy. *Journal of the Geological Society of London*, Vol. 129, 1–37.

— COX, B M, IVIMEY-COOK, H C, and MORTER, A A. 1989. The stratigraphy of the Eriswell Borehole, Suffolk. *British Geological Survey Research Report*, SH/89/2.

BRITISH GEOLOGICAL SURVEY. 1985. *Atlas of onshore sedimentary basins in England and Wales.* (Glasgow: Blackie.)

BRYDONE, R M. 1932. *Uintacrinus* in north Suffolk. *Journal of the Ipswich Natural History Society*, Vol. 1, 158–161.

BULLARD, E C, GASKELL, T F, HARLAND, W B, and KERR-GRANT, C. 1940. Seismic investigations on the Palaeozoic floor of east England. *Philosophical Transactions of the Royal Society*, Vol. 239, A800, 29–94.

CAMERON, T D J, STOKER, M S, and LONG, D. 1987. The history of Quaternary sedimentation in the UK sector of the North Sea Basin. *Journal of the Geological Society of London*, Vol. 144, 43–58.

CARTER, D J, and HART, M B. 1977. Aspects of mid-Cretaceous stratigraphical micropalaeontology. *Bulletin of the British Museum (Natural History), Geology*, Vol. 29, 1-135.

CHROSTON, P N, and SOLA, M A. 1982. Deep boreholes, seismic refraction lines and the interpretation of gravity anomalies in Norfolk. *Journal of the Geological Society of London*, Vol. 139, 255–264.

CLAYTON, K M. 1957. Some aspects of the glacial deposits of Essex. *Proceedings of the Geologists' Association*, Vol. 68, 1–21.

COLE, M J, BROWNING, S M, COOLING, C M, DAVIES, M C, HARVEY, B I, LOVELOCK, P E R, SLADEN, A M, MURRAY, K H, ROBERTSON, A S, and TATE, T K. 1965. Records of wells in the area of New Series One-Inch (Geological) Sudbury (206) Sheet. *Water Supply Papers of the Geological Survey of Great Britain.*

CORNWELL, J D. 1985. Applications of geophysical methods to mapping unconsolidated sediments in East Anglia. *Modern Geology*, Vol. 9, 187–205.

— 1986.   Discussion of 'The location of the Stour buried tunnel-valley using geophysical techniques' by R D BARKER and D HARKER (Quarterly Journal of Engineering Geology, 17, 103–115).   *Quarterly Journal of Engineering Geology,* Vol. 19, 207–208.

COX, F C. 1985.   The tunnel-valleys of Norfolk, East Anglia. *Proceedings of the Geologists' Association,* Vol. 96, 357–369.

— and NICKLESS, E F P. 1972.   Some aspects of the glacial history of central Norfolk.   *Bulletin of the Geological Survey of Great Britain,* No. 42, 79–98.

CURRY, D. 1965.   The Palaeogene beds of south-east England. *Proceedings of the Geologists' Association,* Vol. 76, 151–173.

EDMONDS, E A, and DINHAM, C H. 1965.   Geology of the country around Huntingdon and Biggleswade.   *Memoir of the Geological Survey of Great Britain,* Sheets 187 and 204 (England and Wales).

EHLERS, J, MEYER, K-D, and STEPHAN, H-J. 1984.   The pre-Weichselian glaciations of north-west Europe.   *Quaternary Science Reviews,* Vol. 3, 1–40.

ELLISON, R A. 1983.   Facies distribution in the Woolwich and Reading Beds of the London Basin, England.   *Proceedings of the Geologists' Association,* Vol. 94, 311–319.

— and LAKE, R D. 1986.   Geology of the country around Braintree.   *Memoir of the British Geological Survey,* Sheet 223 (England and Wales).

FUNNELL, B M. 1961.   The Palaeogene and early Pleistocene of Norfolk.   *Transactions of the Norfolk and Norwich Naturalists Society,* Vol. 19, 340–364.

— and WEST, R G. 1977.   Preglacial Pleistocene deposits of East Anglia.   245–265 in *British Quaternary studies.* SHOTTON, F W (editor). (Oxford: Clarendon Press.)

GALE, A S. 1989.   Field meeting at Folkestone Warren, 29th November, 1987.   *Proceedings of the Geologists' Association,* Vol. 100, 73-82.

— and FRIEDRICH, S. 1989.   Occurrence of the ammonite genus *Sharpieceras* in the Lower Cenomanian Chalk Marl of Folkestone.   Appendix pp.80–82 *in* Field meeting at Folkestone Warren, 29th November, 1987.   *Proceedings of the Geologists' Association,* Vol. 100.

GALLOIS, R W, and MORTER, A A. 1982.   The stratigraphy of the Gault of East Anglia.   *Proceedings of the Geologists' Association,* Vol. 93, 351–368.

GIBBARD, P L and ZALASIEWICZ, J A (editors). 1988.   *Pliocene–Middle Pleistocene of East Anglia:field guide.*   (Cambridge: Quaternary Research Association.)

GREGORY, J W. 1898.   Excursion to Sudbury.   *Proceedings of the Geologists' Association,* Vol. 15, 452–456.

— 1915.   The Danbury Gravels.   *Geological Magazine,* Decade 6, Vol. 2, 529–538.

HARMER, F W. 1900.   The Pliocene deposits of the east of England.   II The Crag of Essex (Waltonian) and its relation to that of Suffolk and Norfolk.   *Quarterly Journal of the Geological Society of London,* Vol. 56, 705–738.

— 1902.   A sketch of the later Tertiary history of East Anglia. *Proceedings of the Geologists' Association,* Vol. 17, 416–479.

— 1910.   The glacial deposits of Norfolk and Suffolk. *Transactions of the Norfolk and Norwich Naturalists Society,* Vol. 9, 108–133.

HART, M B. 1973.   Foraminiferal evidence for the age of the Cambridge Greensand.   *Proceedings of the Geologists' Association,* Vol. 84, 65–82.

— BAILEY, H W, CRITTENDEN, S, FLETCHER, B N, PRICE, R, and SWIECICKI, A. 1989.   Cretaceous.   273–371 in *Stratigraphical atlas of fossil foraminifera* (2nd edition). JENKINS, D G and MURRAY, J W (editors). (Chichester: Ellis Horwood.)

HAWKES, L. 1943.   The erratics of the Cambridge Greensand — their nature, provenance and mode of travel. *Quarterly Journal of the Geological Society of London,* Vol. 99, 93–104.

HESTER, S W. 1965.   Stratigraphy and palaeogeography of the Woolwich and Reading Beds.   *Bulletin of the Geological Survey of Great Britain,* No. 23, 117–137.

HEY, R W. 1980.   Equivalents of the Westland Green Gravels in Essex and East Anglia.   *Proceedings of the Geologists' Association,* Vol. 91, 279–290.

HILL, E. 1891.   On wells in west Suffolk Boulder-clay. *Quarterly Journal of the Geological Society of London,* Vol. 47, 585–587.

— 1902.   On the matrix of the Suffolk Chalky Boulder-Clay. *Quarterly Journal of the Geological Society of London,* Vol. 58, 179–182.

— 1910.   Excursion to Sudbury (Suffolk).   *Proceedings of the Geologists' Association,* Vol. 21, 479–482.

— 1912.   The glacial sections around Sudbury (Suffolk). *Quarterly Journal of the Geological Society of London,* Vol. 68, 23–29.

HOPSON, P M. 1981.   The sand and gravel resources of the country around Nayland, Suffolk: description of 1:25 000 resource sheet TL93.   *Mineral Assessment Report Institute of Geological Sciences,* No. 85.

— 1982.   The sand and gravel resources of the country around Sudbury, Suffolk: description of 1:25 000 resource sheet TL84.   *Mineral Assessment Report Institute of Geological Sciences,* No. 118.

JEANS, C V. 1968.   The origin of the montmorillonite of the European Chalk with special reference to the Lower Chalk of England.   *Clay Minerals,* Vol. 7, 311–329.

JEFFERIES, R P S. 1963.   The stratigraphy of the *Actinocamax plenus* subzone (Turonian) in the Anglo-Paris Basin. *Proceedings of the Geologists' Association,* Vol. 74, 1–33.

JUKES-BROWNE, A J. 1903.   On the zones of the Upper Chalk in Suffolk.   *Proceedings of the Geologists' Association,* Vol. 18, 85–94.

KING, C. 1970.   The biostratigraphy of the London Clay in the London district.   *Tertiary Times,* Vol. 1, 13–15.

— 1981.   The stratigraphy of the London Clay and associated deposits.   *Tertiary Research Special Paper,* No. 6.

KNOX, R W O'B. 1990.   Thanetian and early Ypresian chronostratigraphy in south-east England.   *Tertiary Research,* Vol. 11, 57–64.

— and ELLISON, R A. 1979.   A Lower Eocene sequence in SE England.   *Journal of the Geological Society of London,* Vol. 136, 251–253.

— HARLAND, R, and KING, C. 1983.   Dinoflagellate cyst analysis of the basal London Clay of southern England.   *Newsletters on Stratigraphy,* Vol. 12, 71–74.

LAKE, R D, ELLISON, R A, HENSON, M R, and CONWAY, B W. 1986. Geology of the country around Southend and Foulness. *Memoir of the British Geological Survey,* Sheets 258 and 259 (England and Wales).

LAWSON, T E. 1982. *Geological notes and local details for 1:10 000 sheets TM28NW, NE, SW and SE (Harleston, Norfolk).* (Keyworth, Nottingham: Institute of Geological Sciences.)

LINSSER, H. 1968. Transformation of magnetometric data into tectonic maps by digital template analysis. *Geophysical Prospecting*, Vol. 16, 179–207.

MARKS, R J. 1982. The sand and gravel resources of the country around Clare, Suffolk: description of 1:25 000 sheet TL74. *Mineral Assessment Report Institute of Geological Sciences*, No. 97.

— and MERRITT, J W. 1981. The sand and gravel resources of the country north-east of Halstead, Essex: description of 1:25 000 resource sheet TL83. *Mineral Assessment Report Institute of Geological Sciences*, No. 68.

MASSON SMITH, D, HOWELL, P M, ABERNATHY-CLARK, A B D E, and PROCTOR, D W. 1974. The National Gravity Reference Net, 1973. *Professional Paper of the Ordnance Survey, Great Britain*, No. 26.

MATHERS, S J, and ZALASIEWICZ, J A. 1988. The Red Crag and Norwich Crag formations of southern East Anglia. *Proceedings of the Geologists' Association*, Vol. 99, 261–278.

MERRIMAN, R J. 1983. The origin of glauconitic material in Crag deposits from East Anglia. *Proceedings of the Geologists' Association*, Vol. 94, 13–16.

MILLWARD, D. 1980a. Notes and local details for the area around Long Melford and Sudbury (TL84NE, SE) part of 1:50 000 sheet 206 (Sudbury). *Open-File Report of the Institute of Geological Sciences*, No. 1980/4.

— 1980b. Geology of the country around Wixoe and Clare: Explanation of 1:10 560 geological sheet TL74NW, NE, SW, SE. *Open-File Report of the Institute of Geological Sciences*, No. 1980/5.

MITCHELL, G F, PENNY, L F, SHOTTON, F W, and WEST, R G. 1973. A correlation of Quaternary deposits in the British Isles. *Special Report of the Geological Society of London*, No. 4.

MOIR, J A, and HOPWOOD, A T. 1939. Excavations at Brundon, Suffolk (1935–37). *Proceedings of the Prehistoric Society*, New Series, Vol. 5, 1–32.

MOONEY, H, and WETZEL, G. 1957. *The potentials about a point probe in a two-, three- and four-layered earth.* (Minneapolis: University of Minnesota.)

MORTER, A A, and WOOD, C J. 1983. The biostratigraphy of Upper Albian–Lower Cenomanian *Aucellina* in Europe. *Zitteliana*, Vol. 10, 515–529.

MORTIMER, H G. 1967. Some Lower Devonian microfloras from southern Britain. *Review of Palaeobotany and Palynology*, Vol. 1, 95–109.

MORTIMORE, R N. 1986. Stratigraphy of the Upper Cretaceous White Chalk of Sussex. *Proceedings of the Geologists' Association*, Vol. 97, 97–139.

— and WOOD, C J. 1986. The distribution of flint in the English Chalk, with particular reference to the 'Brandon Flint Series' and the high Turonian flint maximum. *In* The scientific study of flint and chert; papers from *The Fourth International Flint Symposium, 1983*, Vol. 1. SIEVEKING, G, and HART, M B (editors). (Cambridge: Cambridge University Press.)

MORTON, A C. 1982. The provenance and diagenesis of Palaeogene sandstones of south-east England as indicated by heavy mineral analysis. *Proceedings of the Geologists' Association*, Vol. 93, 263–274.

MURRAY, K H. 1986. Correlation of electrical resistivity marker bands in the Cenomanian and Turonian Chalk from the London Basin to east Yorkshire. *Report of the British Geological Survey*, Vol. 17, No. 8.

PEAKE N B, and HANCOCK, J M. 1961. The Upper Cretaceous of Norfolk in Larwood, G P and Funnell, B M. The geology of Norfolk. *Transactions of the Norfolk and Norwich Naturalists Society*, Vol. 19, 293–339.

PEARMAN, H. 1982. Caves and tunnels in south-east England, Part 4. *Records of the Chelsea Speleological Society*, Vol. 11.

— 1983. Caves and tunnels in south-east England, Part 5. *Records of the Chelsea Speleological Society*, Vol. 13.

PENNING, W H, and JUKES-BROWNE, A J. 1881. Geology of the neighbourhood of Cambridge. *Memoir of the Geological Survey of Great Britain.*

PERRIN, R M S, DAVIES, H, and FYSH, M D. 1973. Lithology of the Chalky Boulder Clay. *Nature, London, Physical Sciences*, Vol. 245, 101–104.

— ROSE, J, and DAVIES, H. 1979. The distribution, variation and origins of pre-Devensian tills in eastern England. *Philosophical Transactions of the Royal Society*, Vol. 287, B1024, 535–570.

PRESTWICH, J. 1850. On the structure of the strata between the London Clay and the Chalk in the London and Hampshire Tertiary systems: Part 1. *Quarterly Journal of the Geological Society of London*, Vol. 6, 252–281.

— 1852. On the structure of the strata between the London Clay and the Chalk in the London and Hampshire Tertiary systems: Part 3, The Thanet Sands. *Quarterly Journal of the Geological Society of London*, Vol. 8, 235–264.

— 1854. On the structure of the strata between the London Clay and the Chalk in the London and Hampshire Tertiary systems: Part 2, The Woolwich and Reading Series. *Quarterly Journal of the Geological Society of London*, Vol. 10, 75–170.

— 1890. On the relation of the Westleton Beds, or pebbly sands of Suffolk, to those of Norfolk and on their extension inland. *Quarterly Journal of the Geological Society of London*, Vol. 46, 84–181.

REED, F R C. 1897. *Geology of Cambridgeshire.* (Cambridge: Cambridge University Press.)

REID, C. 1890. The Pliocene deposits of Britain. *Memoir of the Geological Survey of Great Britain.*

ROBASZYNSKI, F and AMÉDRO, F. 1980. Synthèse biostratigraphique de l'Aptien au Santonien du Boulonnais á partir de sept groupes paléontologiques: foraminifères, nannoplancton, dinoflagellés et macrofaunes. *Revue de Micropaléontologie*, Vol. 22, 195–321.

ROSE, J, and ALLEN, P. 1977. Middle Pleistocene stratigraphy in south-east Suffolk. *Journal of the Geological Society of London*, Vol. 133, 83–102.

— — and HEY, R W. 1976. Middle Pleistocene stratigraphy in southern East Anglia. *Nature, London*, Vol. 263, 492–494.

— KEMP, R A, WHITEMAN, C A, ALLEN, P, and OWEN, N. 1985. The early Anglian Barham Soil of eastern England. 197–229 in *Soils and Quaternary landscape evolution.* BOARDMAN, J (editor). (Chichester: John Wiley and Sons.)

SHOTTON, F W, BANHAM, P H, and BISHOP, W W. 1977. Glacial–interglacial stratigraphy of the Quaternary in midland and eastern England. 267–282 in *British Quaternary studies, recent advances.* SHOTTON, F W (editor). (Oxford: Clarendon Press.)

SMART, J G O, SABINE, P A, and BULLERWELL, W. 1964. The Geological Survey borehole at Canvey Island, Essex. *Bulletin of the Geological Survey of Great Britain*, No. 21, 1–36.

SMITH, W. 1819. *Geological map of Suffolk.* (London: J Cary.)

— 1820. *Geological map of Essex.* (London: J Cary.)

SOLOMON, J D. 1932. On the heavy mineral assemblages of the Great Chalky Boulder-clay and cannon-shot gravels of East Anglia, and their significance. *Geological Magazine*, Vol. 69, 314–320.

SPATH, L F. 1923-1943. The Ammonoidea of the Gault. Parts 1–16. *Monograph of the Palaeontographical Society.*

SPENCER, H E P. 1967. A contribution to the geological history of Suffolk, Part 3, The glacial epochs. *Transactions of the Suffolk Naturalists' Society*, Vol. 13, 366–389.

SZABO, B J, and COLLINS, D. 1975. Ages of fossil bones from British interglacial sites. *Nature, London*, Vol. 258, 680–682.

VACQUIER, V, STREELAND, N C, HENDERSON, R G, and ZEITZ, I. 1957. Interpretation of aeromagnetic maps. *Memoir of the Geological Society of America*, No. 47.

WARREN, S H. 1957. On the early pebble gravels of the Thames basin from the Hertfordshire–Essex border to Clacton-on-Sea. *Geological Magazine*, Vol. 94, 40–46.

WEST, R G. 1961. Vegetational history of the early Pleistocene of the Royal Society borehole at Ludham, Norfolk. *Proceedings of the Royal Society*, Vol. B155, 437–453.

— and DONNER, J J. 1956. The glaciations of East Anglia and the East Midlands: a differentiation based on stone orientation measurements of the tills. *Quarterly Journal of the Geological Society of London*, Vol. 122, 69–91.

WHITAKER, W. 1872. The geology of the London Basin, Part 1. *Memoir of the Geological Survey of Great Britain.*

— 1885. The geology of the country around Ipswich, Hadleigh, and Felixstowe. *Memoir of the Geological Survey of Great Britain.*

— 1906. The water supply of Suffolk from underground sources. *Memoir of the Geological Survey of Great Britain.*

— BENNETT, F J, and BLAKE, J H. 1881. The geology of the neighbourhood of Stowmarket. *Memoir of the Geological Survey of Great Britain.*

— PENNING, W H, DALTON, W H, and BENNETT, F J. 1878. The geology of the north-west part of Essex and the north-east part of Herts, with parts of Cambridgeshire and Suffolk. *Memoir of the Geological Survey of Great Britain.*

— and THRESH, J C. 1916. The water supply of Essex from underground sources. *Memoir of the Geological Survey of Great Britain.*

WILKINSON, I P. 1988. Ostracoda across the Albian/Cenomanian (Cretaceous) boundary in Cambridgeshire and western Suffolk (eastern England). 1029–1045 *in* Evolutionary biology of Ostracoda, its fundamentals and applications. HANAI, T, IKEYA, N, and ISHIZAKI, K (editors). *Developments in Palaeontology and Stratigraphy*, 11.

WILLIAMS, R B G. 1987. Frost weathered mantles on the Chalk. 127–133 in *Periglacial processes and landforms in Britain and Iceland.* BOARDMAN, J (editor). (Cambridge: Cambridge University Press.)

WILSON, D, and LAKE, R D. 1983. Field meeting to north Essex and west Suffolk. *Proceedings of the Geologists' Association*, Vol. 94, 75–79.

WOOD, C J. 1988. The stratigraphy of the Chalk of Norwich. *Bulletin of the Geological Society of Norfolk*, Vol. 38, 3–120.

WOOD, S V (junior). 1880. The newer Pliocene period in England. *Quarterly Journal of the Geological Society of London*, Vol. 36, 457–528.

WOODLAND, A W. 1946. Water supply from underground sources of the Cambridge–Ipswich district. *Wartime Water Supply Pamphlet. Geological Survey of Great Britain*, No. 20, parts 3 and 10.

— 1970. The buried tunnel-valleys of East Anglia. *Proceedings of the Yorkshire Geological Society*, Vol. 37, 521–578.

WOOLDRIDGE, S W. 1923. The minor structures of the London Basin. *Proceedings of the Geologists' Association*, Vol. 34, 175–193.

— 1926. The structural evolution of the London Basin. *Proceedings of the Geologists' Association*, Vol. 37, 162–196.

WORSSAM, B C, and TAYLOR, J H. 1969. Geology of the country around Cambridge. *Memoir of the Geological Survey of Great Britain.*

WRAY, D S. 1990. The petrology of clay-rich beds in the Turonian (Upper Cretaceous) of the Anglo-Paris Basin. Unpublished PhD thesis, City of London Polytechnic.

— and GALE, A S. *In press.* Geochemical correlation of marl bands in Turonian chalks of the Anglo–Paris Basin. In *Special Publication of the Geological Society of London: High Resolution Stratigraphy.* KIDD, R V, and HAILWOOD, E A (editors).

WYMER, J J. 1985. *The Palaeolithic sites of East Anglia.* (Norwich: Geo Books.)

YOUNG, B, and LAKE, R D. 1988. Geology of the country around Brighton and Worthing. *Memoir of the British Geological Survey* (England and Wales, sheets 318 and 333).

APPENDIX 1

Key boreholes and sections

A BOREHOLES

* indicates boreholes from which BGS holds core and/or samples
† indicates boreholes for which BGS holds geophysical logs
The numbers in brackets after each borehole name, e.g.
(TL 74 NE/15), are those of the BGS records system.

* † Clare Borehole (TL74NE/15) [7834 4536] depth 264.74 m
    River Terrace Deposits, fill of buried channel (Till, Glacial Sand and Gravel, Glacial Silt), Middle Chalk, Lower Chalk (including the Cambridge Greensand), Gault, Silurian.

   Borehole at Old Rectory, Cavendish (TL84NW/10) [8067 4641] depth 137.77 m
   River Terrace Deposits, fill of buried channel (Till, Glacial Sand and Gravel), Middle Chalk.

   Glemsford Borehole (TL84NW/13) [8307 4658] depth 158.50 m
   River Terrace Deposits, fill of buried channel (Till, Glacial Sand and Gravel), Middle Chalk.

† Kettlebaston Borehole (TL94NE/30) [9544 4996] depth 100 m
   Alluvium, Till, Crag, Upper Chalk.

* Hitcham Borehole (TL95SE/17) [9829 5113] depth 36.2 m
   Alluvium, Till, Kesgrave Sands and Gravels, Crag, Upper Chalk.

† Kedington Borehole (TL74NW/29) [7060 4592] depth 110 m
   River Terrace Deposits, Upper Chalk, Middle Chalk.

† Frog's Hall, Lavenham Borehole (TL95SW/20 [9185 5017] depth 126 m
   Glacial Sand and Gravel/Till, Upper Chalk, Middle Chalk.

   Great Cornard (Kedington Hill) Borehole (TL83NE/27) [8899 3863] depth 21 m
   London Clay, Woolwich and Reading Beds, Thanet Beds, Upper Chalk.

* Bulmer Tye (Butler's Hall Farm) Borehole (TL83NW/28) [8340 3791] depth 20.1 m
   Head, London Clay, Woolwich and Reading Beds.

B SECTIONS

Edwardstone gravel pit (Lynn's Hall Quarry) [933 434]: Till, Kesgrave Sands and Gravels, Crag.

Victoria Pit, Sudbury [878 417]: Till, Glacial Sand and Gravel, Crag, Thanet Beds, Upper Chalk.

Cornard brick pit [8885 3843]: Lacustrine deposits, Glacial Sand and Gravel, Till, ?Kesgrave Sands and Gravels, Woolwich and Reading Beds, Thanet Beds, Upper Chalk.

Bulmer brick pit [833 382]: London Clay

Ballingdon chalk pit [8609 4055]: Thanet Beds, Upper Chalk.

Brundon gravel pit [863 417]: River Terrace Deposits.

Bear's Pit, Acton [884 461]: Till, Glacial Sand and Gravel.

Cockfield Quarry [896 559]: Glacial Sand and Gravel.

Old quarry at Swingleton Green, Monks Eleigh [9662 4717]: Crag, Upper Chalk.

APPENDIX 2

BGS photographs

Copies of these photographs may be seen in the library of the British Geological Survey, Keyworth, Nottingham NG12 5GG and at the BGS Information Desk at the Geological Museum, Exhibition Road, South Kensington, London SW7 2DE. They all belong to series A and may be supplied variously as black and white or colour prints and transparencies at a fixed tariff. The photographs 4004–4020 were taken by J Rhodes in 1927. The remainder were taken, during the resurvey, by C C Jeffrey (12521–13004), H J Evans (13241–13265) and B Starbuck (14833–14849).

4004	Glacial Sand and Gravel cutting down into Thanet Beds. Victoria pit, Sudbury.
4005	Glacial Sand and Gravel cutting down into Thanet Beds. Jordan's chalk pit, Sudbury.
4006	As above.
4007	Bedded Glacial Silt on the side of a basin-shaped hollow. California Brickworks, Sudbury.
4008	Chalky till passing down into laminated Glacial Silt. Alexandra Brickworks, Sudbury.
4009	Red Crag sands and disturbed glacial beds resting on Thanet Beds. Sudbury.
4010	Well-bedded Glacial Sand and Gravel. Newton Road sand pit, Sudbury.
4011	Chalk succeeded by Thanet Sands, unfossiliferous Red Crag(?) sands, Glacial Sand and Gravel, and till. Whorlow's chalk pit, east of Sudbury.
4012	Platform of Thanet Beds, Glacial Sand and Gravel etc. Whorlow's chalk pit, east of Sudbury.
4013	'Large boulder of *remanié* chalk included in grey glacial 'brickearth'.' (Calcareous tufa in lacustrine deposits). Cornard brick pit.
4014	Glacial sands etc. resting on Chalk. Railway chalk pit, Sudbury.
4015	Drift cutting down into lower part of Thanet Beds. Railway chalk pit, Sudbury.
4016	Reading Beds and Thanet Beds resting on Chalk. Ballingdon chalk pit.
4017	Sandy basement beds of London Clay. Gestingthorpe Brickyard.
4018	Current-bedded glacial sands. Ballingdon Hill sand pit.
4019	As above.
4020	Chalky till resting on Glacial Sand and Gravel. Ballingdon Grove Brickyard.
12521	Basal London Clay. Bulmer brick pit.
12522	As above.
12541	Wiggery Valley. Gestingthorpe.
12548	Lacustrine deposits at Cornard brick pit.
12549	Small reverse faults in chalky sand and gravel. Cornard brick pit.
12550	Face in lower pit. Cornard brick pit.
12551	Degraded face in upper pit. Cornard brick pit.
12552	London Clay–'Lower London Tertiaries' junction. Cornard brick pit.
12553	Abandoned meander of the River Stour. Cornard.
12554	The Stour Valley from Kedington Hill. Cornard.
12997	Bulmer brick pit.
12998	Bulmer Brick and Tile Works.
12999	As above.
13000	As above.
13001	As above.
13002	As above.
13003	As above.
13004	As above.
13241	Till overlying Glacial Sand and Gravel. Near Stanstead.
13242	Glacial Sand and Gravel. Near Stanstead.
13243	Well-bedded Glacial Sand and Gravel. Near Glemsford.
13244	As above.
13245	Belchamp Walter church and the Belchamp Brook Valley.
13246	Till overlying Glacial Sand and Gravel. Bear's Pit, Acton.
13247	As above.
13248	Banded till. Bear's Pit.
13249	Contorted Sand and Gravel. Bear's Pit.
13250	High Street, Long Melford.
13251	Parish church, Long Melford.
13252	Junction between Thanet Beds and Upper Chalk. Ballingdon, Sudbury.
13253	As above.
13254	Chalk, Thanet Beds and glacial deposits. Victoria Pit, Sudbury.
13255	Parish church, Sudbury.
13256	Thanet Beds. Disused chalk pit, south of Colchester Road, Sudbury.
13257	Thanet Beds, Red Crag and till. Disused chalk pit. As above.
13258	The Old School, Stour Street, Sudbury.
13259	Kesgrave Sands and Gravels outcrop. Great Cornard.
13260	As above, in detail.
13261	Glacial Sand and Gravel. Small gravel pit at Pannell's Ash.
13262	As above.
13263	River Terrace Deposits. Gravel pit south of Brundon.
13264	As above.
13265	Chalk, Thanet Beds, Crag and till. Old pit in Sudbury.
14833	Incised valley in West Suffolk till plateau at Hartest.
14834	Till overlying Kesgrave Sands and Gravels. Edwardstone gravel pit.
14835	Edwardstone gravel pit. Part of face shown in photograph A14834.
14836	Edwardstone gravel pit. Part of the face (enlarged).
14837	Slumping of till, Edwardstone gravel pit.
14838	Edwardstone gravel pit. Till overlying Kesgrave Sands and Gravels.
14839	Slumping of till, Edwardstone gravel pit.
14840	Slumped face, Edwardstone gravel pit.
14841	Edwardstone gravel pit. Part of working face.
14842	As above.
14843	Chalk exposure in River Glem.
14844	As above.
14845	Till with chalk and clay streaks on bank of River Glem.
14846	Reconstituted chalk on bank of River Glem.
14847	As above.
14848	Stream cutting, showing head overlying till. Wickhambrook.
14849	As above.

FOSSIL INDEX

Page numbers of figures and tables are in italics
P after a page number indicates a Plate.

acritarchs   6
Acroloxus lacustris (Linné)   49
Actinocamax verus Miller   12, 23
Ambitisporites cf. *dilutus* (Hoffmeister) Richardson & Lister   6
Ammodiscus cretaceus (Reuss)   *19*
ammonites   14, 15, 20
Anomia sp.   18
Anapholites sp.   14
Anchura (Perissoptera) cf. *maxima* (Price)   14
Ancylus fluviatalis Müller   49
annelids   14
anthozoa   14
Apiculiretusispora cf. *synorea* Richardson & Lister   6
apporrhaids   14
Archaeoglobigerina cretacea (d'Orbigny)   22
Archaeozonotriletes?   6
Arctica sp.   31
Arenobulima advena (Cushman)   *19*
A anglica Cushman   18, *19*
A. cf. *sabulosa* (Chapman)   *19*
A. depressa (Perner)   *19*
Armiger crista (Linné)   49
Astarte sp.   31
Atreta sp.   14
Aucellina   18, *19*
A. coquandiana (d'Orbigny)   14
A. gryphaeoides (J de C Sowerby)   18
A. gryphaeoides morphotype   18
A. uerpmanni Polutoff morphotype   18

Bathyomphalus contortus (Linné)   49
belemnites   23, 44
Birostrina cf. *subsulcata* (Wiltshire)   14
B. concentrica (Parkinson)   *12*, 14
B. concentrica gryphaeoides (J de C Sowerby)   *12*, 14
B. sulcata (Parkinson)   *12*, 12, 14
Bithnyia tentacula (Linné)   49
bivalves   6, 18, 26, 31, 33, 34
Bolivinoides culverensis Barr   23
B. strigillatus (Chapman)   23
bones, mammalian   51, 52
Bourgueticrinus   23
brachiopods   6, 14
   terebratulid   18
Bythoceratina   15

Capillithyris squamosa (Mantell)   *19*, 20
chitinozoa   6
Chondrites   12, 18
Cibicides beaumontianus (d'Orbigny)   22, 23

coccoliths   15, 31
Cribratina cylindracea (Chapman)   *19*
Cymbula? cf. *phaseolina* (Michelin)   14
cysts, dinoflagellate   24, 31
Cythereis (R.) bemerodensis Bertram & Kemper   15
Cytherelloidea globosa Kaye   15
Cytodaria sp.   31

diatoms   26
dinoflagellate cysts   24, 31
Dorothia gradata (Berthelin)   *19*

Eiffellithus turrisseiffeli (Deflandre)   14
Entolium orbiculare (J Sowerby)   14
Entolium sp.   18
Epipholites sp.   14
Euhoplites aff. *trapezoidalis* Spath   14
E. alphalautus Spath   14
E. cf. *subcrenatus* Spath   14
E. inornatus Sapth   14

Flourensina intermedia Ten Dam   18, *19*
foraminifera   15, 18, 22, 23, 26, 28, 31, 44, 49

gastropods   6, 26, 33
Gaudryina austinana Cushman   *19*
Gavelinella ammonoides (Reuss)   22
G. baltica Brotzen   *19*
G. cenomanica (Brotzen)   18, *19*
G. cf. *clementiana sensu* Bailey   22
G. cristata brotzeni (Goel)   22
G. intermedia (Berthelin)   *19*
G. stelligera (Marie)   22
Globigerinelloides rowei (Barr)   23
Globotruncana bulloides Vogler   23
G. linneiana (d'Orbigny)   22
Glyphaea willetti (Woodward)   18
Gonioteuthis granulataquadrata (Stolley)   23
Grasirhynchia grasiana (d'Orbigny)   *19*
Gryphaea   48
Gryphaeostrea canaliculata (J Sowerby)   23
Gyroidinoides nitidus (Reuss)   18

?Holaster sp.   14
Hedbergella amabilis Loeblich & Tappan   *19*
H. brittonensis Loeblich & Tappan   *19*
H. delrioensis (Carsey)   18, *19*
Hemiaster ex gr. *griepenkerli* Von Strambeck-*morrisi* Forbes   18
Hysteroceras aff. *orbignyi* Spath   14
H. carinatum Spath   14
H. cf. *binum* (J. Sowerby)   14
H. cf. *orbygni* Spath   14
H. cf. *varicosum* (J de Sowerby)   14

Idiohamites cf. *tuberculatus* (J Sowerby)   14
inoceramids   15, 23
Inoceramus cf. *anglicus* Woods   14
'Inoceramus' crippsi Mantell   18, *19*
'Inoceramus' ex gr. *anglicus* Woods-*comancheanus* Cragin   18, *19*

'Inoceramus' ex gr. *virgatus* Schlüter   *19*
Inoceramus lissa (Seeley)   *12*, 14
Isocrinus legeri (Replin)   14

Kingena cf. *arenosa* d'Archiac   18

Lamna appendiculata Agassiz   23
Leoniella carminae Cromer   6
Lepidospongia?   22–23
Lima hoperi Mantell   23
Lingulogavelinella arnagerensis Solakius   22
L. cf. *vombensis* Brotzen   22
L. jarzevae Vasilenko   *19*
Loxostomum eleyi (Cushman)   22
Ludbrookia sp.   14
Lymnaea peregra (Müller)   49

Marssonella ozawai Cushman   *19*, 20
Marsupites calyx plates   23
M. testudinarius (Schlotheim)   23
Mimachlamys cretosa (Defrance)   23
miospores   6
molluscs   52
Monticlarella sp.   18, *19*
Mortoniceras (Deiradoceras) cf. *devonense* Spath   14
M. (Mortoniceras) cf. *pricei* Spath   14
Myxas glutinosa (Müller)   49

nannoconids   14
nannoplankton   24, 25, 31
nautiloids, orthocone   6
Neocythere (Physocythere) steghausi (Mertens)   18
Neoflabellina suturalis (Cushman)   22
Neohibolites minimus Miller   12, 14
N. oxycaudatus Spaeth   14
Nerineopsis cf. *coxi* Abbass   14
Nucula (Pectinucula) pectinata J Sowerby   14

Orbirhynchia sp.   *19*, 20
Osangularia whitei (Brotzen)   22, 23
ostracods   6, 18, 49
Ostrea incurva Nilsson   23
Ostrea semiplana J de C Sowerby   23
Oxyrhina sp.   23

Palaega carteri Woodward   18
Pisidium cf. *subtruncatum* Malm   49
P. henslowanum (Sheppard)   49
P. hibernicum Westerlund   49
P. milium Held   49
P. nitidum Jenyns   49
Plagiostoma globosum J de C Sowerby   18
plants   28
Platyceramus?   23
Plectina mariae (Franke)   18, *19*
Porosphaera globularis (Phillips)   23
proto-*Eiffellithus*   14
Pseudoperna boucheroni (*sensu* Woods *non* Coquand)   23
Pseudotextulariella cretosa (Cushman)   18, *19*
Punctum pygmaeum (Draparnaud)   49
Pupilla muscorum (Linné)   52

Pycnodonte (Phygraea) aff. *vesicularis*
(Lamarck)   14
P. (Phygraea) sp.   14
P. sp.   23
P. vesicularis (Lamarck)   23

Quinqueloculina antiqua (Franke)   *19*

radiolaria   26
Retusotriletes cf. *warringtonii* Richardson &
Lister   6
Reussella kelleri Vasilenko   23
R. szajnochae praecursor De Klasz &
Knipscheer   22, 23
Rhizopoterion   23
Rotalipora apenninica (Renz)   *19*
Rugoglobigerina cf. *pilula* Belford   23

?Scapanorhynchus subulatus (Agassiz)   23
scaphopods   14

Schloenbachia   *19*
S. varians (J Sowerby)   20
serpulids   14
sharks' teeth   23
Sphaerium corneum? (Linné)   49
Spiroplectammina praelonga (Reuss)   *19*
Spiroloculina papyracea Burrows,
Sherborn & Bailey   *19*
S. dutempleanus d'Orbigny   23
S. spp. ex gr. *latus* (J Sowerby)   23
Spondylus cf. *latus* or *dutempleanus*   23
sponges, hexactinellid   22
Stensioeina exsculpta exsculpta (Reuss)
22, 23
S. exsculpta exsculpta-gracilis (transitional)
22
S. granulata granulata (Olbertz)   22
S. granulata incondita Koch   22, 23
S. granulata perfecta Koch   22, 23
S. granulata polonica Witwicka   22

terebratulids, large   *19*
Terebratulina striatula (Mantell)   23
Tristix excavatus (Reuss)   *19*
Tritaxia macfadyeni Cushman   *19*
T. pyramidata Reuss   18, *19*
Trochamminoides sp.   26

Uintacrinus   23

Valvata cristata (Müller)   49
V. piscinalis (Müller)   49
Vertigo substriata (Jeffreys)   49

GENERAL INDEX

Figures and Tables in italics
P after a page number indicates a Plate.

Abbas Hall, Great Cornard   32
Acanthoceras rhotomagense Zone   *17*, 20
Actinocamax quadratus Zone   23
Acton   *1*, *24*, 24, *25*, 47, 48, *57*
aegerine   27
aeolian deposits   49
aeromagnetic anomalies   6, *7*, 35
aeromagnetic data   4, *35*, 36
aggregate resources   55
Albian Stage   5, *12*, 12, *13*, 14, 15
Alexandra brickworks, Sudbury   57, 66
alluvium   *34*, 40, 44, 51, 52–53
Alpheton   *1*
Alpheton Borehole   *16*, *21*, 22
ammonite subzones   *13*
ammonite zones   *13*
amphibole   27
Anglian glacial stage   4, 5, 38, 49, 51
   glacial sand and gravel   45–48
Anglian Water Authority   58
apatite   27
Apectodinium hyperacanthum Zone   31
artefacts   4, 51
   Palaeolithic   52
Ashen House Farm Borehole   *59*
Assington   *1*, 29, 42, 47
Aycliff Borehole   20

Ball, Dr K C   18
Ballingdon   15, *16*, 26P, 27, 28, 45, 51
Ballingdon chalk pit   65, 66
Ballingdon Grove   57
Ballingdon Grove Brickyard   66
Ballingdon Hill sand pit   66
Barham Sands and Gravels   4, 37, *38*,
   38, 46, 47, 56–57, *57*
Barham 'sol lessivé'   37
Barnwell Hard Band   14
Barrington   15
basalt   48
basement, magnetic   35
Baythorn Lodge   *17*, 22
Baythorne End Borehole   *16*, *21*
Bear's Pit, Acton   44, 46P, 47, 48, 65, 66
Beestonian cold stage   4, 5, 37
Belchamp Brook   *1*, 2, 15, 28, 39, 50,
   51, 66
Belchamp Otten   *1*
Belchamp Walter   28, 47, 53, 66
Belle Tout Marls   *17*, *21*, 22
bentonites   31
biotite   31
Birdbrook   37, 44
Blakenham Till   39
boreholes
   Alpheton   *16*, *21*, 22

Anglian Water   20
Ashen House Farm   *59*
Aycliff   20
Baythorne End   *16*, *21*
Bower Hall   *16*, *21*
Boxted   *16*, 22
Boyton Hall   *59*
Brettenham   *59*
Bulmer Tye (Butler's Hall Farm)   31,
   65
Canvey Island   8, 10, 36
Castlings Heath   29
Clare   4, 6, *8*, *9*, *10*, 12, *13*, 14, 15, *16*,
   17, 18, *19*, 20, *35*, 39, 48, 65
Claredown Farm   39, *59*
Cranmore Green   *16*, *21*, 21
Duxford   *13*
Ely–Ouse No. 14   12, *13*
Four Ashes   10, 12, *13*
Frog's Hall   *16*, *21*, 22, 65
Glemsford   65
Great Cornard   27, 65
Guilford Kapwood Ltd   *59*
Hadleigh   30
Harwich   6, 36
High Point, Stradishall   *16*
Hill Farm, Boxford   *59*
Hitcham   33, 65
Kedington   *16*, 20, 22, 65
Kettlebaston   *16*, *21*, 65
Lakenheath   10
Little Chishill   *13*
Mill Farm   *16*, *21*
Mundford 'C'   *13*
Popsbridge   22
Rodbridge Corner   *16*, 22
Soham   10
Stafford Works, Long Melford   *59*
Stock   30
Stoke by Clare   *16*, *21*
Stone Street, Boxford   29, *59*
Stowlangtoft   6, *10*, 21, *35*, 36
Stutton   6, 36
Weeley   6, 36
Wixoe   *16*, 20, *21*, 21
Bottom Bed   27, 28
Bouguer gravity anomalies   6, *7*, *9*,
   10–11, 35, 36
Bouguer gravity data   35
Bower Hall Borehole   *16*, *21*
Box, R   *1*, 2, 24, 39, 41, 42, 51, 53
Box valley   28, 30, *34*, 48
Boxford   *1*, *3*, 24, 29, 32, 37, 41, 48, 53,
   57
Boxted   52
Boxted Borehole   *16*, 22
Boyton Hall Borehole   *59*
Brad, R   *1*, 15, 39, 51, 53
Bramertonian age   33
Brandon Flint Series   21–22
Brent Eleigh   *1*, 22, *34*
Brett, R   *1*, 2, 51, 53
Brettenham   *1*, *3*, 37, 42, 44, 59
Brettenham Borehole   *59*
brick clay   57–58
Brickearth   48
brickmaking   48, 52

Bridge Farm   51
Bridge House, Liston   53
Bridge Street   37, 53
Brundon   51, 52, 66
   archaeological site   4
Brundon gravel pit   65
Brundon west   57
building stone   54P
Bullhead Bed   25–26, 27, *50*
Bulmer   *1*, 22, 24, 28, 30, 45, 48
Bulmer brick pit   30, 31P, 57, 65, 66
Bulmer Tye   57
Bulmer Tye (Butler's Hall Farm)
   Borehole   31, 65
buried channels/valleys   37, *43*, 58
   till thickness (isopachytes)   *42*
   see also tunnel-valleys
Bury St Edmunds district   6

calcareous tufa   49, *50*, 66
calcrete   46P
Caledonian orogeny   4, 6
Caledonian structures   35
California brickworks, Sudbury   57
   Glacial Silt   66
Callihoplites auritus Subzone   14
Cambridge Greensand   14, 15–18, *17*,
   18, *19*, 65
Campanian Stage   *17*, 23
Canvey Island Borehole   *8*, 10, 36
Carstone   12
Cast Bed   20
Castlings Heath Borehole   29
Cavendish   *1*, 33, *41*, 51, 52
Cenomanian Stage   5, 15, *17*, 18, 20
Chad Brook   *1*, 2, *50*, 52
Chalk Marl   15, *17*, 18, *19*
chalk quarries   54, 59
chalk rafts   15, 22, 39
Chalk Rock   15, *17*, 20, *21*, 21, 22
   structure contours (Stour valley)   *16*,
   36
Chalk sea   5
Chalk, the   1, 2, 3, 15–23, 24, 27, 28, 29,
   33, 36, 58
   Ballingdon chalk pit   66
   biostratigraphical classification   *19*
   borehole waters   *59*
   Cambridge Greensand   15–18
   Cambridge Greensand to Middle
      Chalk base   18–20
   classification of   *17*
   Lower Chalk   *8*, 15, 18, 20, 65
   Middle Chalk   *3*, *8*, 15, 18, 20, 41, 65
   Upper Chalk   *3*, 20–23, 26P, *34*
Chalky Boulder Clay   1, 39, 43, 44, *50*
channels, drift-filled   39–41, *41*
   see also tunnel-valleys
chert   37
Chillesford Sand Member   33
Chilton Street   57
chlorite   31, 43
churches, flint-built   2, 54P
Clare   *1*, 2, *3*, 6, 36, 37, *41*, 48, 51, 52
Clare Borehole   4, 6, *9*, *10*, 12, *13*, 14,
   15, *16*, *17*, 18, *19*, 20, *35*, 39, 48, 65
   geophysical logs   *8*

Claredown Farm Borehole 39, *59*
clays
  glauconitic 28, 51
  laminated 39
  mottled 27, 28
  see also silts and clays
Cliffe Marshes Borehole 6
Cockfield *1, 3*, 37, 44, 48, 50, *57*
Cockfield Quarry 65
Colchester Green Farm, Cockfield 44
Colchester Road, Sudbury 66
Colne, R *1*, 2, 39, 53, 58
Colne Valley 50
conglomerate, flint 33
Coniacian Stage *17*, 21–22
Cornard 66
Cornard brick pit 29P, *50*, 65
  remanié chalk 66
Cornard Mere 52
Cotton Hall 52
Cowlinge *1, 3*, 58
Crag 1, 2, *3*, 4, 5, 33–34, *34*, 36, 37, 50,
  55–56, 58, 59
Cranmore Green 34
Cranmore Green Borehole 16, *21*, 21
Creeting Till 39
Cretaceous Period 5
Cretaceous rocks 11, 12–23, 36
Cromer Till 43
Cromerian interglacial 5
Cuckoo Tye 26
Cythereis (Rehacythereis) luermannae
  hannoverana ostracod Zone 15

D. cristatum Subzone 14
Danbury Gravels 4
density logs *8, 9*
Department of the Environment 55
Devensian Glacial 5, 39, 49, 52
Devonian rocks 1, 4, 6, *8, 9, 10*, 35
  density values for 10
Devonian sub-crop *35*
dixoni limestones 20
dolerite 44
Douvilleiceras mammillatum Zone 12
Downtonian Epoch 10
drift deposits
  Anglian 38–48
  post-Anglian 49–53
  pre-Anglian 37
Duxford 12
Duxford Borehole *13*

Echinocorys depressula Subzone 23
economic geology 54–60
  brick clay 57–58
  chalk 54–55
  hydrogeology and water supply 58–59
  sand and gravel 55–57
Edwardstone 37, 44, 45P, 53, *57*
Edwardstone Gravel Pit *34*, 65, 66
Elm Tree Farm 33
Ely–Ouse No. 14 Borehole 12, *13*
Ely–Ouse to Essex Transfer Scheme 58
Emsian stage 10
Eocene and Palaeocene rocks 24–32
epidote 27

erratics 4, 43–44, 51
  Chalk 43, 44
  Cretaceous 47P
  flint 43, 44
  glacial 15
  Jurassic 43, 44, 47P
  quartz porphyry 47P
Euhoplites nitidus Subzone 14

faults 10, 33, *34, 35*, 35–36, 66
feldspar 28, 31
Flandrian Interglacial 49
flint buildings 54P, 54
flint conglomerate 33
flints 21, 22, 26P, 27, 28
flow till 39, 44
fluvioglacial gravels, Anglian 4
foraminiferal zones *19*
fossils, Jurassic and Cretaceous,
  reworked 49
Four Ashes Borehole 10, 12, *13*
Foxearth *1*, 28
Frog's Hall Borehole *16, 21*, 22, 65

Gainsborough 2
Gallows Hill 48
gamma-ray logs 6, *8, 9*
garnet 27
Gault *8*, 10, *12*, 12–14, *13*, 15, *19*, 65
geophysical logs 6, *8*, 22
geophysical surveys 6, 39, 59
Gestingthorpe *1*, 24, 28, 30, 45, 48, 57
Gestingthorpe Brickyard, London Clay
  66
Gipping, R 2
Glacial Sand and Gravel 4, *34*, 37, *38*,
  40, *41*, 44, 45–48, 49, 50, 51, 53, 55,
  57, 66
Glacial Silt *38*, 40, 41, 44, 48, 57
  California brickworks 66
glaciation 4, 49
  late-phase 48
  Pleistocene 38
glaciotectonics 48, 49
glauconite 15, 18, 21, 25, 26P, 27, 28,
  31, 51
Glem, R *1*, 2, 15, *16*, 22, 39, 51, 66
Glem valley 50, 52
Glemsford *1*, 2, 20, 37, *41*, 47P, 48, 51,
  52, 66
Glemsford Borehole 65
Goldingham Hall *16, 17*, 22, 28
Gonioteuthis quadrata Zone 23
Gravelgate Farm, Hundon 44
gravity surveys 39, *41*
Great Cornard 32, 51, 52, 54, *57*, 58,
  59
  Kesgrave Sands and Gravels 66
Great Cornard Borehole 27, 65
  Guilford Kapwood Ltd *59*
Great Waldingfield *1*, 42, 47, 53
Great Wratting *58*
Great Yeldham *1, 3*, 39, 41, 48, 53, *57*,
  58
Grime's Graves Marl *17*, 20
ground instability 59–60
groundwater 58, 59

Hadleigh Borehole 30
Hagmore Green *57*
hardgrounds 15
Hartest *1*, 2P, *3*, 37, 39, 52, 66
Hartest Brook 44
Harwich Borehole 6, 36
Harwich Member 30, 31
Haverhill 2
Hawkedon 22, 57
Hawkedon Brook *16*, 22
Head 28, 30P, 41, 49, 49–51, *50*, 50, 51,
  52, 57–58
heavy minerals 4, 25, 27, 43
Highpoint, Stradishall Borehole *16*
Hightown Green *1*
Hill Farm, Boxford, Borehole *59*
Hitcham Borehole 33, 65
Hitcham House 33
Hoplites dentatus Zone 12
hornblende 27
Hoxnian Interglacial 49
Hundon *1, 58*, 59
Hunstanton 18, 20
hydrogeology, and water supply 58–59
Hysteroceras orbignyi Subzone 14
Hysteroceras varicosum Zone 14

illite 28, 31
illite-smectite 27
Industrial Minerals Assessment Unit 55
Inoceramus beds 18, *19*, 20
Ipswichian Interglacial 49, 51, 52

Jordan's chalk pit, Sudbury 66
Jurassic Period 4, 57

kaolinite 28, 31, 43
Kedington *1, 3*, 22, 28, 33, 39, 54
Kedington Borehole *16*, 20, 22, 65
Kedington Hill 51
Kennett, R 2
Kesgrave Sands and Gravels 1, 4, 5, 33,
  34, 37, 44, 45, 46, 47, *50*, 53, 55, 56,
  57
  Great Cornard 66
Kettlebaston *1*
Kettlebaston Basin 33
Kettlebaston Borehole *16, 21*, 65
Kettlebaston Fault *34*, 36
Kiln Farm 48, 50
King's Hill 59
kyanite 27

Lyelliceras lyelli Subzone 12
lacustrine deposits 49, 66
Lakenheath Borehole 10
landslips 53, 60
Lark, R 2, 39
Lavenham *1*, 2, *3*, 22, 33, *34*, 35, 36,
  37, 50, 51
Lawshall 42
Levalloisian artefacts 52
Leymeriella tardefurcata Zone 12
Leys Farm 48
lignite 20
Liston 39, 48, 53
Little Chishill 12

Little Chishill Borehole   *13*
Little Cornard   28, 49, 52, 58
Little Farm   51
Little Waldingfield   36
Llandovery strata   6
lodgement till   *38, 39,* 44
London Basin   25, 31
London Basin syncline   36
London Clay   1, *3,* 3–4, *4,* 5, 24, *25,* 28, 29–32, *34, 50,* 50, 53, 57, 60
  'A' division   30
  Gestingthorpe Brickyard   66
  Harwich Member   30, 31
London Platform   4, 5, 6, 10, 12, *13*
Long Melford   *1,* 2, 28, 34, 39, *41,* 44, 48, 50, 51, 53, 54P, 57, 66
Lower Chalk   15, 18, 20, 65
Lower Chalk/Chalk Marl   *8*
Lower Cretaceous, Gault   *8.* 10, *12,* 12–14, *13,* 15, *19,* 65
Lower Farm   53
Lower London Tertiaries   1, *3,* 4, 24, 28–29, 57, 58, 59
Lower Marl   *17*
Lower Orbirhynchia Band   20
Lower Palaeozoic rocks   6
Lowestoft Till   39, 42, 43, 44
Ludfordian Stage   6
Ludham   4
Ludhamian rocks   33
Lynn's Hall Quarry   37, 65

macrofossil zones, Chalk   *19*
Maldon Court   59
man, early   4, 5
Mantelliceras dixoni Zone   *17,* 20
M. mantelli Zone   15, *17,* 18
M. saxbii Subzone   20
marine deposits
  Pleistocene   4, 5
  Pliocene   4
  see also Crag
Marsupites testudinarius Zone   *17,* 22, 23
Maw   4
Melbourn Rock   15, *17,* 20
Methwold Marl   20, *21*
mica   32, 43, 56
Micraster coranguinum Zone   *16,* 22
M. coranguinum-Uintacrinus socialis zones   *16*
M. cortestudinarium Zone   15, 17, 21, 22
microgranite, porphyritic   44
Middle Chalk   *3, 8,* 15, 18, 20, 41, 65
Middleton   27, 28
Middleton Hall   45
Milden   *24,* 24, 25
Mildenhall   15
Mill Farm Borehole   *16, 21*
Miocene Epoch   5
Monk's Eleigh   *1, 16, 17,* 23, 34, 39, 54, 65
Mortoniceras inflatum zone   14
mottled clays   27, 28
Mount Ephraim Marl   *17,* 20, *21*
Mundford 'C' Borehole   *13*
muscovite   28, 31

National Rivers Authority   *59*
Nedging   *16, 17,* 23
Neostlingoceras carcitanense Subzone   15, 18
Newton   42, 47
Newton Road   *57,* 59
Newton Road sand pit   66
nodules
  phosphatic   12, 14, 15, 18, 33
  septarian   44
Norfolk coast   15
North Sea Basin   5, 25
'North Sea Drift'   4
Norwich Crag   33, 37

Offaster pilula Zone   15, *16, 17,* 23, 24
Old Pit, Baythorn Lodge   *16*
Oldhaven Beds   24, 30

Palaeocene and Eocene rocks   3, 24–32
Palaeocene–Eocene boundary   25
Palaeogene strata   4, 5, 24, 28
  contours on base of   *24*
  isopachyte map   *25*
palaeosols   5, 37
Palaeozoic rocks   6
Pannell's Ash   66
peat   52, 53
pebbles, phosphatic   20
periglacial conditions   49
Permian Period   4
Peyton Hall   41
phosphate   56
phosphatic nodules   12, 14, 15, 18, 33
phosphatic spicules   31
photographs, BGS   66
Pilgrim's Walk Marl   *17,* 20, *21*
Planileberis scrobicularis ostracod Subzone   15
Pleistocene marine deposits   4, 5
Plenus Marls   20
Pliocene marine deposits   4
Pliocene–Pleistocene, Crag   33–34
Popsbridge Borehole   22
Porcellanous Beds   18, *19*
Poslingford   37
Pot Kiln School   59
pre-Cretaceous rocks   6–11
Pre-Ludhamian rocks   33
pre-Silurian rocks   6
Přídolí Epoch   6, 10
  see also Downtonian Epoch
proto-Thames, R   1, 4
'putty chalk'   15

quartz-feldspar porphyry   47P, 48

Railway chalk pit, Sudbury   66
Reading Beds *see* Woolwich and Reading Beds
Reading type beds   27, 28
Red Crag   33, 34, 66
Rede   *1,* 58
remanié chalk   49
  Cornard brick pit   66
resistivity logs   *3, 8, 9, 17,* 22, 39
  Anglian Water boreholes   20

Middle and Upper Chalk   *21*
Ridgewell   *1,* 44
river terrace deposits   5, 40, 49, 51–52, 55, 57
Rob's Farm, Cavendish   37
rockhead contours   *40*
Rodbridge Corner Borehole   *16,* 22
rutile   27

sand and gravel operations, major   *57*
sand
  glauconitic   25
  windblown   5
sand and gravel   1–2, 4, 5, 37, 38, 39, 44, 45–48, 52, 53, 55–57, 59
  Anglian   42
  Beestonian   4
  classification of   *57*
  glacial   *38,* 40, *41,* 44, 49, 50, 51, 53
Santonian Stage   *17*
  Upper Chalk   22–23
sarsens   28, 44
seismic refraction surveys   9
seismic velocity, Devonian rocks   10
septarian nodules   44
Shalford Meadow   53
Shimpling   48
Shoreham Marls   *17, 21,* 22
silts and clays, laminated   *41,* 48, 49, *50*
Silurian rocks   1, 4, 6–9, *8, 9, 10,* 10, 36, 65
Silurian sub-crop   *35*
smectite   28, 31, 43
Smeetham Hall   48
Soham Borehole   10
'sol lessivé'   37
solifluxion   5, 39, 49, 50
Somerton Hall   44, 52
sonic logs   6, *8, 9*
Stafford Works, Long Melford, Borehole   *59*
Stambourne   *1*
Stanningfield   *1,* 39
Stansfield   *1,* 33
Stanstead   34, 47, 48, 66
Starston Till   39, 44
staurolite   27
Stensioeina granulata polonica Foraminiferal Zone   22
Sternotaxis plana Zone   *17,* 21
Stock Borehole   30
Stoke by Clare   51
Stoke by Clare Borehole   *16, 21, 41*
Stoke Row   *57*
Stoliczkaia dispar Zone   14
Stone Street, Boxford, Borehole   29, *59*
Stour Augmentation Groundwater Scheme   58
Stour buried channel/valley   15, 37
  see also Stour tunnel-valley
Stour, R   1, 2, 5, 15, 24, 42, 57, 58
  at Cornard   66
Stour Street, Sudbury   66
Stour tunnel-valley   40, *41*
Stour valley   20, *21,* 22, 25, 26, 28, 29, 30, 39, 44, 45, 48, 49, 51, 52

south side (structure contours)   *16,* 36
Stowlangtoft   15, 20, 22
Stowlangtoft Borehole   6, *10,* 21, 35, 36
Stradbroke–Sudbury Basin   33
structure contours, Chalk Rock Stour valley   *16,* 36
structures   6, 35–36
Sturmer   *1*
Stutton Borehole   6, 36
Sudbury   *1,* 2, *3,* 3, 11, 22, 24, 25, 27, 28, 30, 34, 37, 42, 44, 48, 51, 52, 53, 54, *58,* 59
Alexandra and California brickworks 57
Suffolk Pebble Bed   30
Swingleton Green, Monk's Eleigh   *16,* *17,* 23, 65

Terebratulina lata Zone   *17*
Thanet Beds   1, 5, 24, *25,* 26, 27, 28, 33, *34,* 34, *50,* 66
Thanetian stage   *25*
Thurnian rocks   33
till   1, 2P, 4, 5, 15, 22, *34,* 37, 38, 39, 40, *41,* 41, 42–45, 48, 50, 57, 59
Anglian   39
banded   44, 45P, 46P
contours on base   *43*
flow   39, 44
lodgement   *38,* 39, 44

thickness of, in buried channels   *42*
Whorlow's chalk pit   66
Top Rock   15, *17,* *21,* 21, 22
Totternhoe Stone   15, *17,* 18, *19*
tourmaline   27
Triassic Period   4, 10–11
tunnel-valleys   36, 39, 40, *41,* 42, 47, 48
Anglian age infill   48
Turk's Hall, Boxford   32
Turonian Stage   *17*
Turonian–Coniacian Stages, Lower Chalk   21–22
Turrilites costatus Subzone   20
Twin Marls   20, *21*

Uintacrinus socialis Zone   *17,* 22, 23, 24
Uintacrinus socialis-Marsupites testudinarius zones   *16*
Upper Chalk   *3,* 26P, *34*
Upper Cretaceous *see* Chalk, the

vein quartz   44
velocity logs *see* sonic logs
Victoria Pit, Sudbury   *16,* *17,* 22, 27, 65
Glacial Sand and Gravel   66
volcanic ash bands/layers   30, 31, 35
volcanic fall-out   20
volcanigenic material   26

Ware   6
Ware–Cliffe Marshes line   6, 10

Water Hall   52
water supply   58–59
weathering, periglacial   15
Weeley Borehole   6, 36
Wells Hall, Brent Eleigh   *16,* 22
Wenlock strata   6
Wenner configuration, deep resistivity soundings   10
West Tofts Marl   *21,* 22
Westland Green Gravels   4
Westland Green Member   37
see also Kesgrave Sands and Gravels
Westleton Beds   4
White Ballast   4
Whorlow's chalk pit, till   66
Wickhambrook   *1*
Wiggery Valley, Gestingthorpe   66
Wiggery Wood   48
Wixoe   39, *41,* 51, 52, *57,* *58,* 59
Wixoe Borehole   *16,* 20, *21,* 21
Wolstonian Glacial   4, 38, 39, 49, 51, 52
Woolwich and Reading Beds   1, 4, 5, 24, *25,* 27–28, 29P, 31, *34,* 48, *50*
see also Lower London Tertiaries
Woolwich type beds   27

Ypresian stage   *25*

zeolites   27
zircon   27, 31

BRITISH GEOLOGICAL SURVEY

Keyworth, Nottingham NG12 5GG
(0602) 363100

Murchison House, West Mains Road, Edinburgh
EH9 3LA   031-667 1000

London Information Office, Natural History Museum
Earth Galleries, Exhibition Road, London SW7 2DE
071-589 4090

The full range of Survey publications is available through the Sales Desks at Keyworth and at Murchison House, Edinburgh, and in the BGS London Information Office in the Natural History Museum Earth Galleries. The adjacent bookshop stocks the more popular books for sale over the counter. Most BGS books and reports are listed in HMSO's Sectional List 45, and can be bought from HMSO and through HMSO agents and retailers. Maps are listed in the BGS Map Catalogue, and can be bought BGS approved stockists and agents as well as direct from BGS.

The British Geological Survey carries out the geological survey of Great Britain and Northern Ireland (the latter as an agency service for the government of Northern Ireland), and of the surrounding continental shelf, as well as its basic research projects. It also undertakes programmes of British technical aid in geology in developing countries as arranged by the Overseas Development Administration.

The British Geological Survey is a component body of the Natural Environment Research Council.

HMSO publications are available from:

HMSO Publications Centre
(Mail, fax and telephone orders only)
PO Box 276, London SW8 5DT
Telephone orders 071-873 9090
General enquiries 071-873 0011
Queueing system in operation for both numbers
Fax orders 071-873 8200

HMSO Bookshops
49 High Holborn, London WC1V 6HB
(counter service only)
071-873 0011   Fax 071-873 8200
258 Broad Street, Birmingham B1 2HE
021-643 3740   Fax 021-643 6510
33 Wine Street, Bristol BS1 2BQ
0272-264306   Fax 0272-294515
9 Princess Street, Manchester M60 8AS
061-834 7201   Fax 061-833 0634
16 Arthur Street, Belfast BT1 4GD
0232-238451   Fax 0232-235401
71 Lothian Road, Edinburgh EH3 9AZ
031-228 4181   Fax 031-229 2734

HMSO's Accredited Agents
(see Yellow Pages)

And through good booksellers